ALL

ARE

LOCAL

ALL DATA ARE LOCAL

Yanni Alexander Loukissas

Foreword by
Geoffrey C. Bowker

THINKING CRITICALLY IN A DATA-DRIVEN SOCIETY

THE MIT PRESS
CAMBRIDGE, MASSACHUSETTS
LONDON, ENGLAND

This book was set in PF DIN Pro and DejaVu Sans Mono by The MIT Press. Printed and bound in the United States of America.

Library of Congress Cataloging-in-Publication Data

Names: Loukissas, Yanni A. (Yanni Alexander), author.
Title: All data are local : thinking critically in a data-driven society / Yanni Alexander Loukissas; foreword by Geoffrey C. Bowker
Description: Cambridge, MA : The MIT Press, [2019] | Includes bibliographical references and index.
Identifiers: LCCN 2018030570 | ISBN 9780262039666 (hardcover : alk. paper) 9780262545174 (paperback)
Subjects: LCSH: Electronic information resource literacy. | Media literacy.
Classification: LCC ZA4065 .L68 2019 | DDC 025.042--dc23 LC record available at https://lccn.loc.gov/2018030570

10 9 8 7 6 5 4 3

To Kate

CONTENTS

Foreword ix
Preface xi
Acknowledgments xvii

INTRODUCTION 1

1 LOCAL ORIGINS 13
2 A PLACE FOR PLANT DATA 27
3 COLLECTING INFRASTRUCTURES 55
4 NEWSWORTHY ALGORITHMS 95
5 MARKET, PLACE, INTERFACE 123
6 MODELS OF LOCAL PRACTICE 161
7 LOCAL ENDS 189

Notes 197
Bibliography 219
Index 239

FOREWORD

We've been building our world in both the notional West and global North for about the past two hundred years around the collection and analysis of data—from the natural history and population censuses of yore, to the vast proliferation of data acquisition and analysis practices today. In general, the *West* isn't just the west and the *global North* is not just the north, but historical generalizations have a way of collapsing locals into empires; it's a convenient way of organizing knowledge. In this marvelous volume, Yanni Alexander Loukissas demonstrates that it's turtles all the way down: at whatever level you take an ordered set of givens about the world, you find local practice and exception.

The invocation at the end of the book is a clarion cry for our times: "Do not mistake the availability of data as permission to remain at a distance." It does get messy when you tie data to a place; there were, as Loukissas tellingly shows, over a thousand designations of data in the New York Public Library. You can't simply ingest such data and assume that you can produce scientific facts. All you have access to is data that are machine recognizable as data, and there is a huge amount of work in making it recognizable. One might think of an example from the census: if I fit easily into machine-readable categories, I am easy to count (and therefore my presence counts for something), whereas if I am mixed race and gender nonspecific, I just won't be counted without a lot of extra work. I have served time (yes, it is a sort of prison sentence) looking at biodiversity data. Here Loukissas's insistence on the locality and heterogeneity of data ring true. Most biodiversity data are data from within a hundred miles of an arterial road (it's easier to get to). Global maps of biodiversity work best for areas where most collecting is done by appropriately trained taxonomists, and they are indexed by the specific schools that the taxonomists came out of (a map of fossil specimens in Europe in the nineteenth century was a good map of the Austro-Hungarian empire—folks trained out of Vienna—and British one—folks trained out of Kew Gardens).

Loukissas suggests what for me is precisely the appropriate response: we must create counterdata to challenge normative algorithms. This raises the question of where the site of politics is today. It's hard to think, in the era of Donald Trump, that politics are contained in a Habermasian sphere of rational discourse. In an arena conjured by our data doubles exploiting our every weakness (why does Amazon keep suggesting light stuff that will not ever fulfill me but that will gain my attention?) and magnifying our fears (why does populism become the natural response to induced tribalism?), we are just not collectively performing as rational actors. The central issue is that data about us and the world are circulating much faster than we can have control over. How many of us manage our cookies or read our end-user license agreements carefully? Data are where it's at, and this book provides the best propaedeutic to a reasoned, effective plan of action.

Geoffrey C. Bowker
University of California at Irvine
July 2018

PREFACE

<?XML VERSION="1.0" ENCODING="UTF-8"?> <OBJECT XMLNS:MYCUSTXSL="URN:XSLEXTENSIONS" XMLNS:MSXSL="URN:SCHEMAS-MICROSOFT-COM:XSLT"> <OBJECTDATA> <TITLE>WASHINGTON CROSSING THE DELAWARE</TITLE> <ARTIST> <ARTISTNAME>EMANUEL LEUTZ</ARTISTNAME> <ARTISTDATE> AMERICAN, SCHWÄBISCH GMÜND 1816–1868 WASHINGTON, D.C.</ARTISTDATE> <ARTISTROLE>ARTIST</ARTISTROLE> </ARTIST> <LOCATIONSTRING /> <DATED>1851</DATED> <MEDIUM>OIL ON CANVAS</MEDIUM> <OBJECTNUMBER>97.34</OBJECTNUMBER> <OBJECTID>11777</OBJECTID> <CREDITLINE>GIFT OF JOHN STEWART KENNEDY, 1897</CREDITLINE> <CHAT>THIS DEPICTION OF GEORGE WASHINGTON (1732–1799) CROSSING THE DELAWARE RIVER INTO NEW JERSEY TO LAUNCH AN ATTACK ON THE HESSIANS, GERMAN SOLDIERS HIRED BY GREAT BRITAIN ON DECEMBER 25, 1776–A TURNING POINT IN THE REVOLUTIONARY WAR–WAS A GREAT SUCCESS IN BOTH GERMANY, WHERE LEUTZE PAINTED IT, AND AMERICA. ITS POPULARITY LAY CHIEFLY IN THE CHOICE OF SUBJECT, APPEALING AS IT DID TO FLOURISHING NATIONALISM AT MIDCENTURY NOT ONLY IN THOSE TWO COUNTRIES BUT AROUND THE WORLD. THE WORK'S MONUMENTAL SCALE ADDED TO ITS EFFECTIVENESS. DESPITE SOME HISTORICAL INACCURACIES, THE PAINTING REMAINS AN OBJECT OF VENERATION AND IS ONE OF THE BEST-KNOWN AND MOST EXTENSIVELY PUBLISHED IMAGES IN AMERICAN ART.</CHAT> <ROOMCHAT /> </OBJECTDATA> <GALLERYLOCATION> <CASESECTION DATATYPE="VARCHAR" FIELDTYPE="SYSTEM.STRING" /> <SHELF DATATYPE="VARCHAR" FIELDTYPE="SYSTEM.STRING" /> </GALLERYLOCATION> …

***Source*: Metropolitan Museum of Art (excerpt, not full record)**

The ideas in this book began to take shape in 2006, many years before I started writing it. At the time, I was a graduate student at MIT in Cambridge, Massachusetts, but I traveled to New York City on a regular basis to work on an information technology master plan for the American Wing of the Metropolitan Museum of Art. The largest institutional collector in New York, "the Met" sits at the eastern edge of Central Park. It might seem monolithic at the base of its imposing Fifth Avenue entry stairs. But the institution is actually a composite of independently curated collections. Under the umbrella of a major architectural renovation of the spaces that house the American collection, I was contracted by the Met as part of Small Design Firm, an information and interaction design outfit also based in Cambridge.[1] Our scope of work included the design of wayfinding aids, such as label graphics for the artwork and maps to help visitors explore the collection firsthand as well as a series of digital media installations meant to offer a new kind of museum experience. The challenges that I now address, thirteen years later, in *All Data Are Local,* first presented themselves as I considered how visitors might use data to navigate the Met's vast holdings of American art.

The American Wing's "collections data" have been a work in progress since the mid-nineteenth century when the branch was still a separate building in the park. Since that time, almost twenty-five thousand individual objects, ranging in scale from colonial-era teaspoons to an entire room designed by the architect Frank Lloyd Wright, have been cataloged by the staff as data. Those data have served as a resource for generations of curators seeking to either register or uncover answers to everyday questions about the provenance, authorship, taxonomy, label text, or other assorted details of the myriad objects in the collection.

A reader unfamiliar with collections data might think of them as the contents of a spreadsheet: rows for each object in the collection, and columns for various attributes of those objects. But the attribute fields do not simply register commonplace facts about the artwork. Rather, they contain the kinds of locally relevant details that professional curators rely on for their daily work. The attribute column titled "gallery location," for instance, helps curators track where a piece of the collection is being held, even if only for a moment to clean it or snap a new publicity photograph for an upcoming special exhibition. This list of locations is manually updated in real time to reflect the mundane passage of objects from one room to another. Such records are considered vital, for theft is an ongoing concern of the museum staff.

In following with their original purpose as curators' tools, the American Wing's collections data were long held in what the sociologist Erving Goffman would call the "back stage."[2] Indeed, these data were never intended for outsiders' eyes. So when our team first encountered them, the collections data appeared justifiably strange. They had confounding gaps and curiously dated details, such as label text from other eras. Most peculiar of all, many of the visually striking objects in the collection were represented by tiny black-and-white photographs, only of use as identifiers for in-house staff who already knew the objects intimately.

Our master plan established a strategy for translating and, in a few cases, re-creating these data for the "front stage," where visitors could see and interact with them.[3] Parts of the existing data set were inadequate. For example, our digital media designs required the use of recognizable color images for each object in the collection; the existing black-and-white likenesses would not do. The Met agreed to update their photographs, but not without some hesitation, for this was a serious undertaking, both expensive and time consuming.

While in some ways the American Wing data needed more detail before they appeared in front of visitors, in other ways they contained too much. Visitors, for instance, didn't need to know the history of every time an object was removed from its case for a routine dusting; they only needed to know whether or not the object was on display. In other words, the local specificities of the Met's collections data had to be understood and reframed to make those data more broadly accessible as well as meaningful to visitors.

Even as I worked on the museum master plan, I was also completing a doctorate in which my research focused on the social implications of information technologies for professional life. This research spanned domains as varied as architecture, space exploration, nuclear weapons design, and the life sciences. I was in training to study subjects from a "sociotechnical" perspective: an approach in which the technical operation of a system is examined in tandem with the social relations that it creates or preserves.[4]

I was fortunate to train under a group of eminent scholars of science, technology, and society—a field that might be defined by its focus on locality. This field has illustrated how materially based, everyday patterns of work—locally defined within laboratories, field sites, conference rooms, and even living rooms—can explain the success of science and technology and their expansion throughout the world.[5] My early work with these colleagues has since been documented in two books: *Simulation and its Discontents*, a crosscutting collaborative project on information technologies and professional identities, and *Co-designers: Cultures of Computer Simulation in Architecture*, a more focused exploration of related changes in the building professions, based on my own doctoral dissertation.[6]

Despite my skills as a social researcher, I was hired to work on the museum master plan primarily because of my technical abilities. Educated in both computation and design (I also hold a professional degree in architecture), I was well positioned to think about how emerging information technologies could expand the space of the museum into a new virtual dimension. Yet I could not help but see the museum as a social space too—composed of everyday patterns of work that resembled the sites I was studying in grad school—in addition to a space for design. Before long, I decided to confront the social and cultural contexts for data at the museum, believing that it might help our team develop a master plan that worked locally rather than in the abstract for the sake of the curators and their visitors.

My training in sociotechnical research taught me to delve into contexts like the museum through ethnography: an "interpretative science" in search of meaning, practiced through a combination of close observation and interviews.[7] Ethnography requires an immersive venture into the local. On the museum project, these skills helped me develop an intimate understanding of the museum's data as well as rapport with its staff: those who created and maintained the data. The curators were, by necessity, an integral part of our project. My experiences learning about how they organize their work through data, as well as how those practices have changed over time, rank among the most formative of my professional career. But that was only one part of the story of the museum's data.

Another part—how the data might come into use by visitors—was one that the curators could not easily tell. For that, I had to turn to the visitors themselves and other intermediary informants. I eventually brought my questions about data use to the museum guards, hired to mind the galleries and watch over the art. Notwithstanding their characterization by the museum as "security," I could see that these were important members of the American Wing staff who spent considerable time answering questions from visitors and helping them navigate the building's circuitous plan. Moreover, the guards knew better than anyone what visitors do: how they move, where they go, and even why they get lost. The guards proved to be among the best sources of insight about the potential contexts of data use within the American Wing galleries. It also became apparent that they would be mediating visitor interactions with whatever information technologies we put into place.

Unfortunately, the American Wing's curators didn't initially understand my attempts to include the guards in the design process. From the curators' perspective, the guards were not part of the museum's information infrastructure, or at least they were not intended to be. Nevertheless, an unofficial series of interviews with the guards prompted a turning point in my thinking for the project and more broadly. The insights that I gleaned from speaking with these overlooked experts on visitor activity were revelatory and a long time in the making; the guards were happy to be asked about what they knew. Their conceptions of the museum layout and knowledge about visitor practices proved indispensable for the work of putting together our master plan, including the way that we numbered the floors. Because the American Wing was once a separate building, its floors do not line up with the rest of the Met complex. I learned that the layout of the American Wing and its odd relationship to the rest of the museum meant that visitors had trouble orienting themselves using the museum's own maps.

Building on our work with the American Wing's curators as well as its cadre of insightful guards, our team from Small Design proposed and later implemented a variety of public uses for the collections data. One of the most memorable designs involved the presentation of data inside the American Wing's main elevators. The architects of the renovation, Kevin Roche, John Dinkeloo, and Associates LLP, had already designed beautifully detailed glass cabs and elevator shafts to replace the existing ones. In each cab, all but one of the walls was to be transparent, allowing views directly out of the elevator and into the galleries. But this design had an unfortunate limitation: the spaces visible from the cabs would have few objects on display. Our team had the idea to use large data displays in order to make the single opaque wall of each elevator cab into a virtually transparent surface.[8]

Today, more than ten years after we completed the installation, the elevator displays are still in operation. From within the cabs, visitors can see three-dimensional digital representations of each floor, annotated by details from the museum's collections data. But getting the displays right took some tinkering. Our early designs included everything in the collection. Only after many iterations and feedback from the staff did we converge on a more modest design, with carefully chosen elements to represent each floor. This approach resulted in simpler images that don't replace the experience of the collection but rather invite visitors to step off the elevator and see the objects themselves. We made use of the reconstituted collections data as well as the guards' advice on how to orient visitors. The resulting displays show first-person perspective views of the museum layout, not just the data, and highlight a small number of objects that can be used as landmarks for navigation.

Participating in the American Wing project was one of my earliest experiences helping general audiences to see through data. Today, the notion that data might convey transparency, the appearance of looking beyond the boundaries of our material surroundings, is increasingly common. Yet as I learned at the Met, the view through data is always curated. In ways that are often invisible, data and their experience must be carefully composed, if they are to be comprehensible by a broad audience.

Although we may acknowledge that data and their interpretations are the products of narrowly prescribed practices, we still sometimes expect data to reveal everything or simply the truth of the matter. Whether searching through the extensive records of an institution like the Met, comparing items for sale online, or trying to unpack a complex political event, such as the 2016 US election, we imagine data on their own will grant us insight. Data that are encountered in a museum, created for consumer settings, or collected using political polling, however, are not simply facts. They are cultural artifacts, manufactured and presented within contexts that matter. When data do seem to confer transparency, it is because we are shielded from important details about the context of their creation or display.

As of this writing, the displays that we made for the American Wing elevator are still visible. Yet sadly they are no longer being updated with real-time data. Visitors who step into the elevator today are watching a video on a loop, distantly based on our original interactive visualizations. It was painful and disappointing to learn about this change. Nonetheless, it reinforces my current sensibility about data-driven systems: they are locally contingent and even fragile. Designs dependent on data must be maintained and repaired on a regular basis to ensure that they are in sync with changes in the data themselves or the encompassing infrastructure of the place.

Working on projects intended to produce transparency has taught me much about what—beyond the data—goes into creating that illusion. I have learned to confront the locality of data: the ways in which they are shaped by the context of both their creation (think of the black-and-white photographs useful only for curators) and use (think of the conflicting conceptions of the museum revealed by the guards).

I wrote this book to explain what I have gleaned from years of experience working with unruly data sets in a range of settings. My message to the reader can be summed up as follows: you must learn to look *at data*, to investigate how they are made and embedded in the world, before you can look *through data*. Do not take the apparent transparency of data for granted. When confronted with the task of understanding a new data set, thinking locally is thinking critically.

Lessons from my years of practice and many more as an academic researcher have informed the title claim of this book: all data are local. The book sets out not merely to defend this claim but also to demonstrate its implications for how to engage locally with a range of data sources that the reader might encounter in the public realm: a scientific collection, platform for cultural history, archive of the news, and online marketplace for housing.

Many years after signing on to the Met project, I am both a designer and scholar of information. I wrote *All Data Are Local* from the position of this dual allegiance, and my hope is that the book will resonate with colleagues in both fields. For designers, it is a primer on the social lives of data. For scholars, it demonstrates how design can extend and embody the work of sociotechnical studies.[9] But the book is also intended for a more general audience, for whom both data and design might be equally opaque. I believe it can help uninitiated readers begin to think critically about data as well as the design of systems that are data driven.[10] Across scales, from software applications to social media communities to smart cities, critical thinking about data is poised to become the new basis for identifying effective and ethical design.

Yanni Loukissas
Atlanta, Georgia
August 2018

ACKNOWLEDGMENTS

This book was made possible by the support of many people, communities, and institutions. As the final manuscript came together, remembering all those who deserve thanks held an outsize space in my mind. Now, I am eager to externalize those acknowledgements, for there are many.

Let me start with my students. Long before this book went to print, the ideas and examples within were tested by designers, planners, journalists, artists, activists, scientists, and engineers who took my courses at Cornell, Harvard, and Georgia Tech. The members of the Local Data Design Lab at Georgia Tech, however, have been my most consistent interlocutors during the final phases of the writing. They helped me separate the salient insights from my own preoccupations. Some of my students—Krystelle Denis (Harvard) as well as Michelle Partogi, Benjamin Sugar, Christopher Polack, and Peter Polack (all from Georgia Tech)—helped me prepare the visualizations and other presentations of data that appear throughout the book. Further acknowledgments for the specific efforts of each are listed in the image credits and endnotes. I wish to single out Peter Polack for his dedicated work on three different chapters. As I hope the reader will come to appreciate, his thoughtful contributions made the book better throughout. In addition, Eric Corbett and Firaz Peer (also from Georgia Tech) both contributed to the research for chapter 5.

Next, let me acknowledge those who created the settings for my earliest research on the book. At MIT, I would like to thank William Porter and Sherry Turkle as well as the late Edith Ackermann and William Mitchell. Their mentorship set me on a course far from where I began, as an architect with an aptitude for computer modeling. David Mindell, my postdoctoral supervisor, and other members of the MIT Program in Science, Technology, and Society first helped me see how design might be useful for uncovering traces of otherwise-invisible sociotechnical relationships. David Small and his team at the eponymously named information design studio, Small Design Firm—which sadly closed shop during the last few months of my writing—first introduced me to collections data and the collectors behind them.

Many of the ideas in this book were incubated at metaLAB(at)Harvard, where I spent two years (2012–2014) working with some of the sharpest minds on collections: Matthew Battles, Kyle Parry, Robert Pietrusko, Cristoforo Magliozzi, Jesse Shapins, Jeffrey Schnapp, and Jessica Yurkofsky, among others. Matthew, specifically, stoked the fires of my early interest in data as cultural artifacts and later worked with me on the material that formed the basis of chapter 3. Throughout the project, he continued to offer invaluable advice. In parallel, I learned a great deal from the variety of participants in the many data and design workshops that I collaboratively led while researching this book: Beautiful Data I and II, DigitalSTS and Design, Humanities Data Visualization, and Data Walks.

Thank you to my colleagues at Georgia Tech. The faculty and staff in the Program in Digital Media, the School of Literature, Media, and Communication, and throughout Tech encouraged, and in some cases directly supported, my work on a daily basis. My sincere gratitude to the following people: Charles Bennett, Ian Bogost, Kenya Devalia, Carl DiSalvo, Keith Edwards, Jennifer Hirsch, Nassim JafariNaimi, Chris LeDantac, Janet Murray, Elizabeth Mynatt, Laine Nooney, Anne Pollock, Fred Rascoe, Melanie Richard, Juan Carlos Rodriguez, Dean Jacqueline Royster, Michael Terrell, and Richard Utz. Lauren Klein and Gregory Zinman deserve special acknowledgments for they did the grim work of reading my earliest drafts. Their insights have become foundational to the book. Beyond my home institution, I want to thank the members of the Society for Social Studies of Science who provided a necessary academic context for my work on the book, particularly the core members of "digitalSTS": Laura Forlano, Steven Jackson, David Ribes, Daniela Rosner, Hanna Rose Shell, and Janet Vertesi.

Thanks as well to the contributors from my many different field sites. The staff of the Arnold Arboretum, William (Ned) Friedman, Peter Del Tredici, Michael Dosmann, Lisa Pearson, Kyle Port, and Kathryn Richardson, among others, made my work on chapter 2 possible. Thank you to the staff and contributing librarians of the Digital Public Library of America, especially Nate Hill, who enabled my research for chapter 3. Thank you to Johanna Drucker, Sergio Goldenberg, Jacob Eisenstein, and Francis Steen, all of whom provided crucial guidance on the development of chapter 4. Thank you to Tim Franzen, Dan Immergluck, Alison Johnson, Tony Romano, and Housing Justice League and Research/Action members, who informed my work on chapter 5. Furthermore, my sincere thanks to all of the anonymous interview subjects—the archivists, botanists, computer scientists, data scientists, designers, curators, journalists, librarians, organizers, security guards, planners, and relators who I cannot mention by name.

On the production side, the book benefited enormously from editing support by Ada Brunstein and Anna Lee-Popham. My editorial team at the MIT Press, Gita Devi Manaktala, Virginia Crossman, Susan Clark, Cindy Milstein, and others on their staff worked closely with me during the final few months. Marge Encomienda is responsible for the inspired book design. Special thanks to Stefanie Posavec, who allowed us to adapt her design for the cover. I could not have asked for more supportive, insightful, and reliable guides on the last leg of this journey.

Parts of the book have appeared in print elsewhere. Chapter 2 appeared in an earlier form in "A Place for Big Data: Close and Distant Readings of Accessions Data from the Arnold Arboretum," *Big Data and Society* 3, no. 2 (December 1, 2016), https://doi.org/10.1177/2053951716661365. Selections from chapter 3 were included in "Taking Big Data Apart: Local Readings of Composite Media Collections," *Information, Communication, and Society* 20, no. 5 (May 4, 2017): 651–664, https://doi.org/10.1080/1369118X.2016.1211722. A version of chapter 5 appeared in "All the Homes: Zillow and the Operational Context of Data," in *Transforming Digital Worlds. iConference 2018. Lecture Notes in Computer Science, Vol 10766*, ed. Gobinda Chowhury, Julie McLeod, Val Gillet, and Peter Willett (Cham, Switzerland: Springer, 2018), 272–281. Thank you to all the unnamed journal and conference reviewers as well as the editors who supported those efforts to publish my early findings.

I am grateful to Anita Say Chan, Laura Kurgan, Catherine D'Ignazio, and Jer Thorp as well as the anonymous book and proposal reviewers who read the full manuscript at different phases in its development. Sincere thanks as well to Geoffrey Bowker, who graciously agreed to write the foreword.

I would like to acknowledge the funders of this work: Arts and Humanities Research Computing at Harvard University, as well as GVU/iPAT, the School of Literature, Media, and Communication, and the Ivan Allen College at Georgia Tech.

Finally, my family deserves the deepest gratitude. In Atlanta, James Diedrick, Jeffrey Diedrick, Leeanne Richardson, and Karen Sherrard all offered heartfelt advice, emotional support, and childcare when I really needed it. Randy Loukissas in Philadelphia and Jennifer Loukissas and her family in Washington, DC, have been continual champions for this work. Jennifer lent her keen editorial eye to the project. The resulting text is much better because of it. At home, Felix and Sonja—both born during the writing of this book—have been important companions with me on this journey, whether they knew it or not. Their presence gave me the perspective that I needed to focus on the high stakes implications of the book and to finish without too much fussing. Last of all, this book would not have happened without Kate Diedrick, who I met around the time that I first started the research. In the years since, she has had more influence on its scope and sensibility than anyone else.

INTRODUCTION

FROM DATA SETS TO DATA SETTINGS

While reading through the accessions data of Harvard University's Arnold Arboretum, one of the largest and most well-documented living collections of trees, vines, and shrubs in the world, I came across the record for a cherry tree, *Prunus sargentii*, named for its collector, the botanist Charles Sprague Sargent. The data suggest that this specimen was retrieved by Sargent on an expedition to Japan in 1940. Yet Sargent died thirteen years earlier, in 1927. How might we decipher the convoluted origins of this tree: uprooted from Japan and planted in US soil on a timeline that makes little sense to an outsider?

In the collections data of the New York Public Library, I discovered 1,719 different conventions for writing the date (i.e., _ _-_ _-_ _ _ _ is just one). Some of these date formats are strange, some are approximate, and some are in languages other than English. Taken together, they reveal the unexpected diversity of cataloging practices that one institution can contain. Recently, the institution contributed its data to a broad initiative called the Digital Public Library of America (DPLA). Containing data from libraries, museums, and archives across the country, this "mega meta collection" manages a confounding number of conflicting formats.[1] How can we expect to make sense of such heterogeneous sources and draw connections among them?

Querying NewsScape, a real-time television news archive hosted by the University of California at Los Angeles, offers access to more than three hundred thousand broadcasts dating back to the Watergate era—so much data that it cannot be seen independent of the algorithms used to search it. How should we differentiate the substance of news data from the computational procedures, such as natural language processing, necessary to access and analyze them?

The website Zillow, an interface to real estate data, purportedly, on all the homes in the United States, seems to facilitate a new level of transparency for the housing market. I can use the site to track the fluctuating market value of my own house or any one of the more than one hundred million properties listed, most of which are not even for sale or rent. But from within the consumer-centered context that Zillow has created, the effects of the inflated housing market on low-income communities across the country remain invisible. How are we to learn about the hidden impacts of our own uses of data?

These four examples introduce a number of challenges that can arise when trying to make sense of unfamiliar data: contradictions, conflicts, and opacities as well as the unintentional effects of both data collection and use. Yet they reinforce a single point, expressed in the title of this book: all data are local. Indeed, data are cultural artifacts created by people, and their dutiful machines, at a time, in a place, and with the instruments at hand for audiences that are conditioned to receive them.[2]

When I made this observation in 2013 to a roomful of colleagues at the University of California at Berkley in the course of an irreverently titled symposium, The Data Made Me Do It, my words were met with a level of incredulity.[3] But in the ensuing years, we have all become more suspicious of the apparent biases and skewing effects in data. Even major news outlets have begun to report on the dark side of the data revolution, including accusations that Google has inadvertently trained its search algorithms on racist data, that strict measures of scholastic achievement can compel schools to "teach to the tests" or even attempt to cheat them, and that manufactured evidence in so-called fake news might have greatly influenced the 2016 US presidential election.[4] Even academics in the social sciences, who are expected to treat sources with more nuance, are embroiled in debates about how their own data might be unethically skewed by p-hacking: a technique by which researchers artfully manipulate the variables and scope of their analyses in order to produce results that might be considered statistically significant.[5]

A broad range of data-driven professions, which have become accustomed to using evidence collected in distant places and times, are publicly raising questions about how to best handle their most valued sources of information. It is not sufficient to identify and eliminate the most evident biases in data. We must learn to work differently, to uncover the inherent limitations in all such sources, before they lead us further astray.

Today, it is too easy to acquire data sets online without knowing much about their locality—where are they produced and used elsewhere—and how that may matter. We have come to rely on the availability of data as generic resources for reasoning not only in scholarship but in education, politics, industry, and even our personal lives. It is now commonplace for researchers, government institutions, and businesses alike to make their data available online, although often without enough accompanying guidance on how to put those data to good use. The problem starts with our language: the widely used term *data set* implies something discrete, complete, and readily portable. But this is not the case. I contend that we must rethink our terms and habits around public data by learning to analyze data *settings* rather than data *sets*.

This book is an exploration of nuances in data practice long debated in scientific laboratories, libraries, newsrooms, and activist communities, but more recently set aside in the contemporary rush to capitalize on the increasing availability of data.[6] I have found that experienced scientists, librarians, journalists, and activists implicitly know that looking for the local conditions in data can help them to work more effectively, and counter biases when necessary. We rarely need to discard data simply because they are strange. After all, data are useful precisely because they provide unfamiliar perspectives, from other times, places, or standpoints that we would not be able to access otherwise. The strangeness of data is its strength.

One ready example of how to use data locally is already at your fingertips. It is the way you might use this book's index. An index, like a data set, is a collection of related yet discrete expressions (the key terms in the book) gathered into a condensed, accessible reference. If the reader were to flip to the index now, they would find that it is most useful in conjunction with the corresponding source (this text) to which each independent entry refers. On its own, an index serves as little more than a teasingly abstract trace of what could be learned by reading the entire book. Nevertheless, the index is useful, provided the book is also at hand. Too often we attempt to use a given data set as a complete work, such as a book, rather than an index to something greater.

Instead of treating data as independent sources, we should be asking, *Where* do data direct us, and *who* might help us understand their origins as well as their sites of potential impact? The implications of these questions are threefold. For practitioners who want to work with data, understanding local conditions can dispel the dangerous illusion that any data offer what science and technology studies scholar Donna Haraway calls "the view from nowhere."[7] For students and scholars, attention to the local offers an opportunity to compare diverse cultures through the data that they make or use. Finally, local perspectives on data can awaken new forms of social advocacy. For wherever data are used, local communities of producers, users, and even nonusers are affected.

COLLECTIONS AS CASES

This book demonstrates how to understand data settings, not simply data sets, by taking the reader through six principles over an equal number of chapters. Chapter 1 takes on the first principle and the title claim that *all data are local*. The next four principles are illustrated by the concrete cases first introduced at the start of this chapter. They exemplify areas of utmost importance for creating an informed public: science communication, cultural history, journalism, and the housing market.

- The accessions data of the Arnold Arboretum can help us understand, first and foremost, that *data have complex attachments to place*, which invisibly structure their form and interpretation.
- The DPLA can help us see that *data are collected from heterogeneous sources*, each with their own local attachments.
- NewsScape offers an opportunity to learn how *data and algorithms are entangled*, with far-reaching implications for what it may mean to be informed in the future.
- Finally, the case of Zillow shows how *interfaces recontextualize data*, with striking consequences for the value that we place on our homes and those of others.

These cases reflect the challenges of working with publicly available data—challenges that are often overlooked in the abundant and pressing conversations on personal data and privacy.[8] The first two cases explain the local contingencies of data, and how discontinuities among data can lead to conflicts. The next two look at the implications of data's locality for how we might understand higher-level computational structures: first algorithms, and then interfaces.

I use the term *local*, further explained in the next chapter, as a relative designation. Over the course of this book, each case offers an opportunity to incrementally explore and elaborate on what local can mean in relationship to data: from a form of place attachment, exemplified by the accessions data of the Arnold Arboretum, to the traces of such attachments found in the accumulated sources of data infrastructures, such as the DPLA or NewsScape. In the final case, on Zillow, the local is primarily identifiable in negative terms; local details are stripped away from data in order to create the "frictionless" interfaces desired by today's harried users. For their fickle audiences, companies in today's "interface economy" seek to make data accessible and actionable anywhere.[9] In doing so, they both obscure and then supplant the traditional meaning-making power of the local.

Toward its end, this book shifts from theoretical principles to strategies for practice. Chapter 6 leaves the reader with a culminating principle only hinted at above—*data are indexes to local knowledge*—and a set of practical guidelines that build on the preceding cases:

- Look at the data setting, not just the data set
- Make place a part of data presentation
- Take a comparative approach to data analysis
- Challenge normative algorithms using counterdata
- Create interfaces that cause friction
- Use data to build relationships

The book concludes with a question: How can we rework open data initiatives to make data settings versus data sets both accessible and actionable? Keeping this long-term ambition in mind, the reader might approach each case in the book by considering what it takes, beyond simply access, to make data usable effectively and ethically.

Let me now make a caveat: despite the provocatively broad claim on the cover of this book, I do not address *all* types of data. Most of the examples that I use throughout the text can be characterized as collections data. These are data that help people to manage distributed work with large quantities of objects, organisms, texts, images, and more. I focus on collections data for three reasons that I hope will make my argument more accessible to readers.

My choice, first of all, has to do with the concreteness of collections data. They refer to actual subjects in the world: plants, books, broadcasts, homes, and even people. Second, collections data are likely to be familiar to many readers. Social media have turned our lives into vividly documented collections of "friends," "favorites," and "shares." Likewise, e-commerce sites like Amazon are collections. This is partially because, in recent years, standards for collections data have converged with object-oriented approaches to programming—a strategy for defining computational systems in terms of classes of objects and their attributes—in order to produce a powerful model for a broad array of online interactions.[10] Third, collections data have historically been used to do the work of curation (from the Latin *cura*, meaning "care")—a practice that for reasons I will get into later in this chapter, necessitates a local perspective. But when the data that describe large, complex collections are aggregated, without regard to their localities, we can be blinded to important distinctions within data. If unacknowledged, these distinctions can sometimes become structural fissures and even lead to a collapse.

Consider, as a stark example readily available in US public consciousness, the role of data in creating and, at first, obscuring the mortgage crisis of 2007. Several years before the market collapsed, in 2004, an eccentric financial manager named Mike Burry with a knack for identifying unique investment opportunities pored over reams of documents describing home loans that comprised a financial product known as a mortgage bond. At the time, private home mortgages were deemed the most stable kind of investment. Beyond ensuring the American dream of homeownership, the resulting mortgage-backed bond market served as the bedrock of the US economy.

As Burry slowly uncovered, the dream would become a nightmare for many homeowners. These bonds weren't based on uniform home loans with fixed terms. Rather, they were comprised of claims on returns from a heterogeneous reserve, including thousands of independent mortgages with varying risks. Many of them turned out to be "subprime": loans made at alarmingly high, variable interest rates and with a high risk of foreclosure. In order to tease apart the risks that each bond contained and understand the chance that the entire bond could fail, Burry had to work through a lengthy legal and financial prospectus. Back then, he might have been the only person to have done so, apart from the attorneys responsible for its assembly.

Michael Lewis recounts this tale in *The Big Short: Inside the Doomsday Machine.* Lewis's book, later adapted into the Oscar-winning film of the same name, tells of the creation and collapse of the mortgage bond market. At the time, all subprime mortgage bonds were considered equivalent, with their value set and secured by the unimpeachable ratings agencies, Moody's and Standard & Poor's.[11] Each mortgage bond represented innumerable pieces of loans that remained largely unexamined by Wall Street. Bonds based entirely on mortgages, explains Lewis, "extended Wall Street into a place it had never before been: the debts of ordinary Americans."[12]

Based on interviews with the few eccentric investors who saw it coming, Lewis's book introduces us to the backroom world of Wall Street where the housing crisis of 2007 began. "The people at Moody's and S&P," notes Lewis, "didn't actually evaluate the individual home loans, or so much as look at them. All they and their models saw, and evaluated, were the general characteristics of loan pools."[13] Meanwhile, the banks presumed that they were passing off any potential liability by repackaging the risk. Also, they strongly suspected that even if the liability did catch up to them, the federal government would bail them out, which ultimately it did, but not before hundreds of thousands of people lost their homes to foreclosure.

By reading between the lines of the mortgage bonds, Burry discovered the contingent nature of each mortgage: its size, interest rate, payment structure, and inherent risk. Moreover, he learned that the number of interest-only, riskier mortgages contained within these bonds was increasing over time. This meant defaults were immanent. Burry leveraged this insight to bet against the housing market so as to "short" the mortgage bonds.

While others dealt blindly with the bonds as aggregates, Burry's research allowed him to see the housing crisis several years before it hit. Unfortunately, rather than using this knowledge to help those most imperiled by these practices, he chose to profit from their effects. *The Big Short* works as a cautionary tale about financial bubbles, but also as a lesson about the locality of data: data have heterogeneous sources, and there are severe implications for those who don't know how to read them with a discerning eye.

Mortgage data, by the way, are collections data too: records on individual entities used to identify and organize them as part of a larger composite. Nevertheless, the principles espoused in *All Data Are Local* can be quite broadly applied, beyond data that deal exclusively with collections. Other types of data, not addressed in this book, are also local and dependent on knowledge about their settings for responsible use. My own varied experiences with data have impressed this on me. In a study of human and machine interactions from the first lunar landing, I learned how Apollo 11 astronauts Neil Armstrong and Buzz Aldrin, aboard the lunar module (nicknamed "Eagle"), were distracted by unexpected and ultimately inconsequential *feedback data* from their guidance computer.[14] The astronauts could not decipher a series of outputs—the values "1201" and "1202"—on their display/keyboard interface. Recognizing them as alarm signals, the astronauts wasted critical seconds reaching out to ground control for help in deciphering these data. In the time that elapsed, the Eagle overshot its landing site and nearly crashed into the surface of the moon. In another study of human-machine communication, this time in a hospital operating room, I witnessed a surgeon, overly reliant on *sensor data* from an electrocardiogram, overlook a pool of blood slowly forming around his sneakers.[15] Another observer in the room warned the surgeon in time to save the patient from bleeding out. In both cases, astronauts and medical professionals were focused on data, and not the broader setting or context.

Time and again, I have encountered such signs of the insistent locality of data across data types. My discussion of collections data, however, is not meant to be comprehensive in scope. I have selected examples that illustrate the limits of universalizing ambitions for data and prompt us to think about how they might be used more conscientiously. The reader might notice that my focus is predominantly on US data. In fact, the cases were chosen specifically because of their proximity and interrelationships. Together, they characterize a particular data-driven society. Although this is a significant limitation to my work, it also presents strategic opportunities. These cases can be used to challenge the unwarranted dominance on the internet of data created in the United States.[16] Seeing how data are local, I argue, can help us put data in their place, materially as well as politically.

LOCAL METHODS AND GOALS

All Data Are Local is assembled from a combination of qualitative findings on data cultures and exploratory data visualizations. Both are informed by extended ethnographic fieldwork, including interviews, workshops, and hands-on engagements with data, conducted over the course of seven years. My use of the term *ethnographic* echoes anthropologist Sherry Ortner's explanation of the method as an "attempt to understand another life world using the self—as much of it as possible—as the instrument of knowing."[17] Indeed, this is a book based largely on my own experiences as an observer and participant in data settings, guided by a desire to understand data through the perspectives and practices of both their keepers and subjects.

My approach is unconventional, but it builds on substantial research in data studies—an area of scholarship that has emerged recently in response to the increasing importance of data in everyday life. Data studies, which seeks to make sense of data from a social and humanistic perspective, has been a significant area of scholarship ever since information scholars Geoffrey Bowker and Susan Leigh Star published *Sorting Things Out: Classification and Its Consequences* almost two decades ago. Their book established the terminology and stakes for thinking about the social lives of data. But the worlds of data look strikingly different today, in 2019, than they did at the end of the twentieth century when *Sorting Things Out* was published.[18] We need new ways of thinking about and looking at the role of data in the public realm.

As explained above, my empirical focus is on four different collections of data. Each chapter documents my efforts to understand one of those collections within its spatial as well as social and technological contexts.[19] These cases might have been a means of reinforcing similar points by tracking one or more themes across many examples. Instead, I take each collection as an opportunity to open up new territory, to ask what each data setting can reveal that is distinct about the locality of data. Moreover, I try to engage these collections reflexively, considering my own position and relationship to the data and their subjects. Each collection is local for me, the investigator, in a different way.

In order to carry out this agenda, I employ a variety of methods for studying data, which I collectively refer to as *local readings*.[20] As the phrase implies, I treat data as texts: cultural expressions subject to interpretative examination. All my readings of data rely on insights gleaned from their keepers, who use their own local knowledge to explain the contingencies of their data, which are not apparent otherwise. Moreover, local reading necessitates examining data comparatively. As cultural anthropologist Clifford Geertz explains, one local condition is most productively understood not in relation to some imagined universal but instead relative to another locality.[21] Sometimes my local readings are made possible by looking at different collections juxtaposed with one another. In other instances, these local readings involve looking at how data are made differently over time, but within the same institution. More experimentally, reading locally can mean imagining how data might be seen in new ways, using speculative yet nevertheless locally imagined modes of visualization.[22] From my perspective, visualization is simply another way of reading data. Each chapter in the book contains one or more visualizations that extend as well as enrich claims made in the text.[23]

My use of visualization is informed by a long history of design practices that produce informative and expressive experiences of data.[24] Today, most writing in the area of data visualization is pragmatic, offering techniques for hands-on work with data. Edward Tufte's book *The Visual Display of Quantitative Information* first introduced many contemporary scholars and practitioners to the potentials, pitfalls, and pleasures of looking at data graphically. Yet Tufte and more recent authors treat data as given.[25] It is time that we learned how to visualize critical thinking on the subject of data.

The visualizations in this book are meant to be exercises in first-person participation and inquiry within data cultures. As such, the visual results might at first appear odd or atypical to the reader. For example, some are entirely textual as opposed to graphical. These visualizations focus on showing the structure and texture of data, rather than offering clear visual patterns, telling stories, or answering narrowly defined questions, as more conventional instances of data visualization might do. In engaging these visualizations, the reader should be ready (as they must with any evidence) to do some of their own interpretative work. Visualizations are, after all, also texts.

One note about the critical sensibilities of this study: my methods are significantly informed by though distinct from those employed by the cohort of scholars who practice under the banner of *critical data studies*.[26] Geographers Rob Kitchin and Tracy Lauriault explain the purpose of this emergent area of investigation:

> To unpack the complex assemblages that produce, circulate, share/sell and utilize data in diverse ways; to chart the diverse work they do and their consequences for how the world is known, governed and lived-in; and to survey the wider landscape of data assemblages and how they interact to form intersecting data products, services and markets and shape policy and regulation.[27]

The work of critical data studies—to unpack, chart, and survey—is typical of critical approaches to scholarship. Across various areas of information studies, the term *critical* has been used to support projects that challenge the status quo: critical games, critical literacy, critical making, and critical design.[28] As a mode of engagement that illuminates biases and assumptions we might otherwise unconsciously adopt, a critical approach is imperative for data studies and practice. But critical reflection has its own limits; it can be detached rather than responsible, analytic rather than affective, or conceptual rather than hands-on.

I take a critical stance, but also explore approaches to working with data that are less distant and cerebral than critical reflection implies. In order to do so, my approach integrates lessons from the feminist ethics of care. Unlike critical reflection, care embraces affect, material engagement, and a host of concerns sometimes invisible in conventional work with technology. Care is critical in that it calls attention to neglected things.[29] But it is more than critical reflection; it is a doing practice.[30] In pursuing opportunities not only for critical reflection on data but in support of care too, I hope to bring largely unrecognized and unrewarded local sensibilities into efforts to understand data.

AGAINST DIGITAL UNIVERSALISM

Finally, a note about the stakes of what I am proposing. The increasing availability of data in public life is part of a broader social and technological transformation: the rise of digital media. Since their early manifestations, digital media have been promoted as a means of independence from local constraints. "Being digital," prophesizes tech visionary Nicholas Negroponte in his 1995 text by the same name, means "less and less dependence upon being in a specific place at a specific time." Eventually, he posits, "the transmission of place itself will start to become possible."[31]

Twenty years later, in 2005, Negroponte launched his "One Laptop per Child" (OLPC) project with the aim of producing a rugged, cheap, and low-power computer, complete with its own open-source software, that might help poor, rural schoolchildren worldwide further their own education by "connecting to the world."[32] Much has been written about OLPC that I will not reprise here.[33] In fact, I would like to put aside questions about whether the project has been successful or not, and instead contemplate OLPC as representative of a broader ideology of place agnosticism for digital media. Negroponte's project, after all, is not designed to improve the places where its young intended users live but rather to make them less dependent on those places. Mike Ananny and Niall Winters, critics of the project, explain that "OLPC sees the child as the agent of change and the network as the mechanism of change."[34] We might ask, Why is it so important for digital media to be place agnostic? Who benefits from claims about the boundlessness of being digital?

The history of our digital media infrastructures is replete with important places. Consider the origins of the World Wide Web, the primary mechanism through which data are made publicly available on the internet today. It began as a local, "home brew" project.[35] The web's tripartite structure—an address system (URI), network protocol (HTTP), and markup language (HTML)—was first developed by Tim Berners-Lee and his colleagues in a specific setting, CERN (the European Particle Physics Laboratory in Geneva, Switzerland), as a primarily textual scientific communication platform for distributed teams of physicists and engineers. Only years later was it reimagined as an infrastructure for the internet: a model for web page cataloging, delivery, and design that might be expanded indefinitely.[36]

The web's uniquely identifiable pages, connected by unidirectional links, might have solved the local problem of information management at CERN, but it makes for an ungainly way of structuring everything that we do on the web today. Moreover, websites have not become the self-contained places that Negroponte imagined might be "transmitted" anywhere. Rather, they are assemblies of data and algorithms that are composed, maintained, and eventually encountered in a variety of settings that matter to their use.[37]

Why, then, do so many creative people working in digital media today embrace what ethnographer Anita Chan calls "the myth of digital universalism"? This ideology, asserts Chan, leads us to falsely believe that despite our varying local circumstances, "once online, all users could be granted the same agencies on a single network, all differences could dissolve, and everyone could be treated alike."[38] Perhaps, operating from within their cloistered innovation centers, digital elites can become heedless to the contingencies of place. Indeed, it is difficult to predict how other conditions, in other places and times, governed by unfamiliar norms, might clash with their own judiciously designed systems. To use a more contemporary example, despite being mired in scandals over the spread of conspiracy theories and hate speech online, prominent leaders in Silicon Valley are still reluctant to acknowledge that they cannot control what users do with their platforms.[39]

One reason universalist aspirations for digital media have thrived is that they manifest the assumptions of an encompassing and rarely questioned free market ideology.[40] If you are not influenced by your setting, you are a more independent and economically rational individual. If your reach can extend indefinitely, your profits can always grow. If you can participate from anywhere, competition is at its strongest. When digital media normalize human behavior and diminish local effects, the market gains strength.

The market, however, should not be the sole means of evaluating digital media and, by extension, contemporary manifestations of data. The diversity and prosperity of the world's varied and contingent digital practices depend on our acceptance of data's locality. In fact, the stakes for the future of the internet could not be higher. If left unchallenged, digital universalism could become a new kind of colonialism in which practitioners at the "periphery" are made to conform to the expectations of a dominant technological culture.[41]

If digital universalism continues to gain traction, it may yet become a self-fulfilling prophesy by enforcing its own totalizing system of norms. Fortunately, there is still time to halt the march toward placelessness. As I argue in this book, learning to look at the local conditions of data can be a form of resistance to the ideology of digital universalism and threat of erasure that it poses to myriad data cultures.

Resisting digital universalism, and instead seeking out the ways in which data are local requires hard work. But it is work that can begin modestly: by learning about the data-shaping power of settings, such as the Arnold Arboretum; taking account of the inherent discontinuities in practices of data collection, such as those illuminated by the DPLA; acknowledging the limits of algorithms that only recognize normative patterns in the news, as NewsScape demonstrates; and confronting the devastating effects of the data-driven housing market on users and nonusers alike, as exemplified by Zillow. Overlooking the locality of data means being naive to the contradictions, conflicts, opacities, and unintentional impacts produced by efforts to universalize data.

All Data Are Local is meant to stand as an alternative to the disassociated theoretical treatises and practical manuals on contemporary practices with data. The book asks readers to consider how a local perspective can transform practices designed to make sense of data. Readers will learn how to engage with the local conditions of data productively in ways that lead toward just ends for those who make, use, or are themselves the subjects of data. Few of us today do not fall into one or more of these categories.

By the end of this book, I hope the reader will acquire new sensibilities about both data and the local. For I aim to do more than simply proclaim that data are local or muster the evidence necessary to convince readers; indeed, this claim may simply confirm what some data-savvy audiences intuitively know. Rather, this book tackles a pragmatic question: How do local conditions matter for understanding data in everyday practice? In the chapters that follow, I consider that question as it manifests in a range of specific settings.

1

LOCAL ORIGINS

What does it mean to say, "All data are local"? Before proceeding to the main cases of the book, let me further unpack the title claim. It relies on two concepts, *data* and *local*, that are pervasive in contemporary popular and academic writing, but less commonly seen together. In order to explain their pairing for the purposes of my argument, I need to address a number of misconceptions about each term independently. In the process, I will take the opportunity to acknowledge the contributions of other scholars who have laid the substantial foundations for this work. Let me turn to conceptions of data first.

TAKING DATA APART

Data Is Not Singular

Five-thirty-eight, the go-to website in recent years for journalistic commentary on questions about data, proclaims, "It should be 'data is.' It just sounds wrong otherwise."[1] Anyone who has worked sufficiently with data has probably thought this or heard it announced by others. Although contemporary dictionaries and style guides maintain that both singular and plural uses are accepted (yes, "all data are local" is grammatically correct), popular writers and increasingly academics shy away from the plural.[2] Instead, they employ the term as a mass singular noun. After all, data seem ubiquitous, like water, air, or oil—to which data have sometimes been crudely compared as the next great resource. But those words have no singular. Data does: datum. Moreover, every decision about word choice cannot simply be about what sounds right, for that changes over time. If "data are" sounds funny, I ask the reader to embrace its strangeness, treating this as an opportunity to reflect on their use of the term more generally.

To emphasize the relative subtlety of this suggestion, consider that some scholars ask us to put aside the term *data* altogether. In order to foreground the human production of data, digital humanities theorist Johanna Drucker has a radical proposition: as opposed to saying *data*, which in Latin means "given," why not substitute *capta*, meaning "taken."[3] This would serve as a useful reminder; capta must come from somewhere and someone. I don't dispute Drucker's incisive point. Yet *data* is too widely used to be abandoned just now. We can, however, retain the plural usage of the term in order to reinforce a multifaceted perspective. Unlike the subjects of many mass singular nouns, we can take data apart (without the need for a chemical reaction). When we do so, we find that data are heterogeneous in ways that matter.

Throughout the book, I use *data* in its older plural form (unless I am referring to the word itself, which is of course singular). This acknowledges that data arise from and are used in varied circumstances worth acknowledging. Even the largest data sets are often agglomerations of sources made in varying conditions—whether they are collected with different instruments, in particular organizations, or simply at distinct moments. While investigating the DPLA (profiled in chapter 3), which brings together an eclectic collection of digital artifacts and formats from across the United States, I was struck by the diversity of its formats. These differences can be attributed to many things. For instance, while archivists seek to determine and preserve the original or authentic ordering of a data collection, librarians attempt to bring each individual data point into an encompassing ordering system, such as the official Library of Congress notation, Dewey decimal classification, or more flexible Dublin Core, thereby facilitating access for a broader public.

Furthermore, we cannot speak about data independent of the plurality of encompassing structures through which they are collected, encoded, and managed: databases. These structures, typically hidden from view behind user interfaces, vary widely as well.[4] And like data themselves, databases are subject to material and historical constraints.

It is worth noting here what I mean by material constraints. I am building off the notion of "materialities of information representation" described by informatics scholar Paul Dourish. In his recent book *The Stuff of Bits*, Dourish writes at length about the "material forms in which digital data are represented and how these forms influence interpretation and lines of action."[5] So, for example, relational databases—the most common form today—are a 1970s' invention of IBM, meant to be data agnostic. They treat data as generic content, rigidly held in an array of rows and columns. Yet as critical cartography has taught us, there is nothing neutral about imposing a grid on the world.[6] Earlier databases were structured differently. Hierarchical, networked, and attribute-value systems are frequently overlooked alternatives in which data are more entangled with their database structure. Thus, the plural view of data should be extended to the ways in which data are stored, and in turn manipulated and retrieved.

Data Are Not Universal

For the last few decades, scholars have developed empirical accounts of how data vary from one scientific or engineering context to the next.[7] These studies have largely sought to complicate a widely held but simplistic perspective: that data are universal and invariable. Science and technology studies (STS) scholar Bruno Latour deftly captured this purified conception of data with the term *inscription*. Latour explains inscriptions (i.e., data) as "objects" created for the production of scientific arguments, "which have the properties of being mobile but also immutable, presentable, readable and combinable with one another."[8] As a shorthand for this collection of features, Latour calls inscriptions "immutable mobiles."[9]

Many have challenged this overly abstract definition of data by exposing the ways in which data practices and data themselves differ from one context to the next. Research on the diversity of data has been conducted in the subdiscipline of laboratory studies, which Latour helped to pioneer, as well as in museums, health care, space exploration, and climate debates, to name a few.[10] Outside of scientific and engineering work, studies of data reveal even further variation. Conceptions of data differ enormously within social and humanistic research—the latter of which has only recently adopted the term at all.[11] Each disciplinary community has its own rituals for determining when something should count as data. Moreover, in different arenas of public or civic life, data and infrastructures to support them are multiplying rapidly. In *The Data Revolution*, a comprehensive look at emergent practices with data, Rob Kitchin takes pains to account for all the sorts of data one might encounter today: "Data vary by form (qualitative or quantitative), structure (structured, semi-structured, or unstructured), source (captured, derived, exhaust, transient), producer (primary, secondary, tertiary), and type (indexical, attribute, metadata)."[12]

But accounting for the various genres of data does not help us understand their use and meaning. For as critical theorist Jonathan Culler reminds us, "Meaning is context-bound, and context is boundless."[13] Accounts of data too often leave out this broader context. Curiously, Latour's term *inscription*, translated from the Latin roughly as "write into," suggests that data are embedded, rather than autonomous.[14] Indeed, the term itself—if not Latour's usage of it—resonates with my view of data as enmeshed in sociotechnical contexts that shape their production and interpretation. How can we learn to engage with this encompassing context, which Kitchin, writing alongside Tracey Lauriault, calls a "data assemblage"?[15]

The Arnold Arboretum, profiled in chapter 2, illustrates how data are entangled within a knowledge system and inscribed in a place. For instance, the arboretum has long collected data on "provenance type," a classification used to distinguish plants by their origins: "W" marks a plant collected in the wild; "C" marks a plant sourced from a nursery or other institutionalized setting; "U" marks a plant of unknown origin; and "Z" singles out a plant as a cutting, taken from a wild plant that is already part of the arboretum collection. Knowing why provenance type matters to researchers at the arboretum, how it figures into the troubled history of the institution, and what about these designations incites controversy today is essential to understanding the arboretum's data along with the questions that they can and cannot help answer. Outside the context of their supporting knowledge system, such data are misleading at best.

Data Are Never "Big"

Big data began as a buzzword, created by industry and embraced by popular media, in order to describe data sets of the 1990s and early 2000s so massive that they outgrew the existing tools to manage them.[16] More recently, big data has gained legitimacy as an independent area of research. Academics have since tried to formalize the term. Kitchin and coauthor Gavin McArdle define big data as those data that are principally high magnitude in volume, variety, and velocity. Yet a range of other characteristics are often present in data sets described as big: exhaustive, high resolution, relational, and flexible.[17]

I have come to find, however, that more than anything else, *big data* is a term that speaks to our contemporary feelings about data. Social researcher Kate Crawford has called big data, data at the scale that inspires anxiety.[18] Her reference point for high anxiety data is the trove of intel collected from social media by the likes of the US National Security Agency or Britain's Government Communications Headquarters—purportedly to identify the next terrorist plot.

Those agencies have amassed enough data for all of us to feel surveilled, while also containing sufficient gaps to sow unshakable doubt among intelligence agents themselves. "We don't have context!" agonize the creators of the innocently named "Squeaky Dolphin" project, a recently exposed big data initiative created by the Government Communications Headquarters. Anxieties proliferate on all sides of big data. Crawford's definition is amusing and apt, particularly because "big" is relative. It can mean different things to different communities. Each group that makes use of data has its own capacities and thresholds for being overwhelmed by "the anxieties of big data."[19]

Despite these useful metacharacterizations of big data, all digital data are by definition agglomerations of small, discrete signals, represented as 0s and 1s in computer memory. Each byte, each item in a list, and each row or cell in a spreadsheet is part of a composite with a complex provenance. Whether generated by algorithms, created by instruments, or keyed in by catalogers, digital data have their own contingencies that are useful to understand. Frequently these differences, called "inhomogeneities" by historian of science and technology Paul Edwards, go unnoticed. Or if they are noted, they are filtered out in order to create the illusion of consistency, necessary to perform large scale queries or calculations with a data set.[20]

The term *big data* has endured beyond many expectations. Moreover, the ideology of big data has pervaded existing data initiatives that would not be considered big when judged by their scale alone. In examples such as the Arnold Arboretum, DPLA, NewsScape, and Zillow, the ideology of big data has infiltrated workaday practices with data sets that measure only in the tens of thousands or millions of entries. Aspiring to the ideology of big data means seeking to collect everything on a subject, downplaying the importance of data's origins, and assuming that data alone can entirely supplant other ways of knowing.[21]

Returning to the example of the DPLA, considered to be big data within the organization (signaling the library's anxiety about it), we can see that data from distributed collections across the country do not simply add up. As mentioned in the previous chapter, in one contributing collection alone, I found more than a thousand different date formats, from "ca. _ _ _ _s" (of which there are 640 instances) to "probably before _ _ _ _" (of which there are only 7). Speaking about these data as "big" effaces the complexity of what initiatives such as the DPLA bring together.

Data Are More than Rhetorical

Although I advocate returning to an older plural usage of *data*, my understanding of the term is not entirely old school. Indeed, I depart from historical understandings of data as primarily rhetorical. In its earliest applications as a theological expression in the seventeenth century, data meant either "the set of principles accepted as the basis of an argument" or "facts, particularly those taken from scripture."[22] In the late eighteenth century, the term was adapted to describe the results of empirical observations and experiments of the kind now associated with scientific practice. Even today, in varied realms of scholarship, data often means "alleged evidence."[23]

Information studies scholar Christine Borgman writes that "entities become data only when someone uses them as evidence of a phenomenon, and the same entities can be evidence of multiple phenomena."[24] She offers the example of excerpts from an old family photo album. One researcher might treat these as evidence of particular clothing styles common to an era. Another may use them to verify family relations. In other words, things become data within interpretative acts. Although this may hold true in some scholarship, I am focused on a broader spectrum of uses for data and their emergent social roles in the public realm. Today data are part of functional sociotechnical systems from which they cannot be easily separated. In this sense, data can be operational rather than rhetorical.

The operational function of data is most evident in the case of housing, explained in chapter 4. Zillow's interface draws together data from public and private sources across the United States in an effort to not simply represent the state of the housing market but also reshape its dynamics. The site offers visitors new estimates every day on homes across the country—their own or prospective purchases and rentals. In doing so, Zillow is effectively establishing a new subject position for buyers and sellers in real estate: seemingly empowered by numbers, but actually blinkered to the broader implications of their own market-frenzied choices for the future of affordable housing options. Data have become operational; they are enmeshed in the practices and politics of everyday life.

Not Everything Can Become Data

In her rhetorical treatment of data, Borgman argues that they are defined not by "what" but rather "when."[25] Anything can become data, observes Borgman, if it is taken up as evidence in an argument, including texts, photographs, and even traces of pigment from an archaeological field site. Making data, though, is not simply an act of naming. In thinking about when something becomes data, we must lend weight to the material processes involved; measuring, recording, and otherwise capturing the world are processes with physical constraints. Making data means bringing a subject into a preexisting system, defined by durable conditions of data collection as well as storage, analysis, and dissemination.[26] Aspects of the original subject are inevitably lost in translation.

Some subjects may not be translatable at all. Chapter 4 will explore the limitations of datafication for the news.

One way to think about the constraints of data is by considering them as "media," for data can be characterized by their material conditions as well as their expressivity, both defining features of media forms.[27] Across the domains of science, cultural history, journalism, and real estate, data constrain how people physically and cognitively interact with the subjects of their interest, whether they are plants, books, news stories, or properties.[28]

Thus, scholars of media can aid in our understanding of how data function in the world. For instance, media theorist and designer Lev Manovich usefully juxtaposes the database with the narrative—an older system for cultural expression—by focusing on their structural differences. In fact, he argues that the database has deposed narrative to become the dominant media form of our time. "As a cultural form," explains Manovich, "database represents the world as a list of items, which it refuses to order." A narrative, meanwhile, "creates a cause-and-effect trajectory of seemingly unordered items (events)."[29] Over the course of the last half century, digital media have given rise to a variety of creative experiments, which propose how databases might function symbolically and affectively in relation to narrative. Some of those experiments, related to the themes of the book, are introduced in chapter 6.

Writing more broadly about the transformative potential of "electronic" media, theorist Marshall McLuhan, a well-known figure, if self-contradictory and controversial, might own some responsibility for the derision that the local conditions of digital media currently receive. In 1964, he wrote that "after more than a century of electric technology, we have extended our central nervous system itself in a global embrace, abolishing both space and time as far as our planet is concerned."[30] That made for a dramatic headline in the 60s. I have found that data actually preserve competing measures of space and time, however, effectively reinforcing the importance of those measures. Think of the innumerable date formats of the DPLA that haven't been so easily reconciled.

My understanding of data is plural, embedded, small, operational, and material. This is not the popular view of data, but it is in line with an emergent discourse developing among scholars at the intersection of media and information studies and STS.[31] Moreover, it is a necessary point of departure for the main subject of this book: how conceptions of data might benefit from increased attention to the local.

DELIMITING THE LOCAL

Local Is Not Lesser

The term *local* has become ubiquitous. It is used popularly as a modifier for widespread phenomena, objects, and actions: local food, local time, local anesthesia, or local elections.[32] In the academic world, the term has long been invoked by social scientists to describe knowledge practices grounded in particular places, usually those inhabited by small, indigenous, marginal, or non-Western cultures.[33] Local knowledge explains how different communities make sense of things in their own way. Clifford Geertz once explained local knowledge by posing a question: "Who knows the river better, the hydrologist or the swimmer? Put that way, it clearly depends on what you mean by 'knows.'"[34] As in this example, discourses on local knowledge have often compared science, an allegedly universal knowledge system, with other ways of knowing. But such juxtapositions can echo dangerous, marginalizing descriptions of nonscientific cultures.

I use the word *local* as it has been appropriated into postcolonial and feminist science and technology studies, precisely in order to upend the dichotomy between scientific and othered cultures of knowledge production.[35] As feminist theorist Sandra Harding explains, "Postcolonial histories and studies of contemporary projects have shown that in important respects modern sciences and technologies, no less than other culture's traditions of systematic knowledge, are local knowledge systems."[36] Harding's work helps scholars to see that when modern science and technology are understood in local terms, their histories, present-day controversies, and future limitations become clearer. In recent years, STS scholars have illustrated how all knowledge systems are rooted in practices and politics related to their time and place.[37]

Knowledge systems surrounding data are no exception.[38] In STS, the term *local* is often used interchangeably with *situated*, another word that has served feminist critiques of expert knowledge.[39] Like the word *situated*, *local* is used to explain how knowledge is embodied, mediated, and historically grounded. Yet *situated* is sometimes interpreted as being about social and material conditions exclusively, while local puts more weight on the relevance of place. The concept of place, another widely theorized term, adds an awareness of spatial conditions to the investigation of knowledge practices. As geographer Tim Cresswell remarks, "Place combines the spatial with the social." Unfortunately, many practices with data are imagined to unfold in an infinitely small space. "Our consciousness of place," writes Cresswell, "all but disappears when it appears to be working well."[40] There are a number of ways that dimensions of the local, like place in general, can be suppressed from view.

At the Arnold Arboretum, many ways of knowing coexist: that of the scientists who run experiments at the institution, the arborists who tend to "their data" (the plants), the commuters who use the grounds as a way to cut across the neighborhood to the Forrest Hills train station, the foragers who appear in the spring, and even the participants in religious rituals that are known to take place on the arboretum grounds under cover of darkness. Each of these ways of knowing the arboretum is local in its own way.

Local Is More than a Geolocation

Bringing the term *local* (and *place* by association) into conversations about how data are situated requires significant explanation. For example, space and place must be differentiated. Space is precise, geometric, and geolocated. Place is something less easily defined: contingent on experience, defined by meaning, and susceptible to changing social designations. In his work on "critical regionalism," architectural theorist Kenneth Frampton explains that localism transcends abstract space, and instead hinges on conditions of culture and identity.[41] Moreover, locality is not merely physical. It can be virtual. In the 1980s, novelist William Gibson introduced readers to the potential of cyberspace: not only a space, but a place in which new morphologies and social forms are possible.[42] Scholars such as Michael Benedikt, Manuel Castells, and William Mitchell also theorize the meaning of place in digital life, even to the point of questioning its continued relevance.[43] Mitchell observes that "once there was a time and a place for everything; today, things are increasingly smeared across multiple sites and moments in complex and often indeterminate ways."[44] Castells portrays this as a new set of relations that transcend place. We no longer occupy spaces defined by their place-based meaning, his argument goes. Rather, we exist in a "space of flows," along which data, commodities, and capital move freely.[45] I argue that flows are still shaped by places, although in ways that have not been adequately unpacked.[46]

One of the examples in which the local transcends geolocation is the news—a topic that I address more fully in chapter 4. Local news has long been defined by a set of topics, a language, and a point of view; these dimensions of place cannot be encompassed by a pin on a map. Media and communication studies scholar Christopher Ali explains the complexity of defining local news media: "While localism can be loosely defined as the mandate for broadcasters to be responsive to their communities, localism as a symbolic category means different, often contradictory things to different people at different times."[47]

The news demonstrates just how strange yet important the local can be. Today, networks and newspapers around the United States have replaced much of what might be called "local news" with cheaper material published by national and international sources. This is a problem, for newspapers and television stations that are actually local tend to be more trusted by their audiences.[48] Meanwhile, these larger, less trusted sources are still bound by topics, languages, and points of view as well as conditions such as timing, format, and systems for dissemination that shape their place in the media landscape. Indeed, the phrase *local news* can distract from the reality that all news is local, created in narrow social and historical settings for audiences prepared to receive it. The 2016 US presidential election revealed just how crucial place can be: news produced by profit-making networks based in large cities in the United States was perceived as the propaganda of urban elites, while "fake news" from Russia was blamed for unfairly skewing discussions of the election online.

On a separate note, it is useful to distinguish my focus on the locality of data from the concept of *local data*. In community-based practices, the phrase *local data* already has currency. Often understood as neighborhood scale, nonexpert, or even idiosyncratic, *local data* can mean data made through "artisanal" practices or by some creative individual's "magic hands," rather than clearly documented and replicable processes.[49] This use of this phrase suggests results that are explicitly subjective, personal, and perhaps intentionally nonreplicable. Such forms of data may eschew the use of standards altogether and do not assume that their development is easily extensible. In that sense, this book is not about local data.

Instead, I argue that seemingly impersonal, large-scale data sets are also local. Even the US census, which purports to account for every person in the country, is entangled with local considerations.[50] It uses historically situated racial categories, only recognizes individuals with permanent addresses, and under the current administration, the 2020 census may omit "undocumented" residents who nevertheless make substantial contributions to US economic and cultural life.[51] Most of the data sets used in this book are similarly not the result of so-called artisanal practices but are nevertheless local. Acknowledging that all data are local means understanding that the phrase *local data* is redundant. As with *raw data*—now familiar as an oxymoron thanks to the work of Geoffrey Bowker and media studies scholar Lisa Gitelman, who have shown that data are always already curated or "cooked"—we should stop using the phrase *local data*.[52] Why not simply talk about all data sets in terms of their settings?

Local Operates on Many Scales

The usefulness of the term *local* depends on its relative nature. Every local condition must be defined in its own way. As Geertz explains, "In the solar system, the earth is local; in the galaxy, the solar system is local; and in the universe, the galaxy is local. To a high-energy physicist, the particle world—or zoo—is, well, the world. It's the particle, a thread of vapor in a cloud of droplets, that's local."[53]

Similarly, in computing, *local* indicates the relative placement of a digital file: a folder is local to your hard drive, your hard drive is local in a network, and your network is local on the Internet. In common use, the term *local* gains relevance in relation to national or global contexts.[54] Despite widespread aspirations for everything to have global relevance, the local retains its importance because of the enduring stubbornness of the physical world: we cannot be everywhere at once, and every specific environment is continually changing.

Local conditions vary and operate at different scales. When it comes to housing, for example, there are numerous scales at which data face local contingencies. Houses are formally appraised in value at the scale of the individual unit. Zooming out, neighborhoods tend to rise and fall in value as a whole. At a higher level, counties create assessment models for tax purposes, which are eventually aggregated into sites like Zillow. Meanwhile, cities and states regulate home values. At the federal level, mortgages are secured, making today's inflated market possible. Housing data are shaped by conditions at all of these scales.[55]

Local Coexists with Global

Although all data are locally made, data do not serve exclusively local needs. They must often appeal to nonlocal audiences. In fact, scholars in infrastructure studies see the local and global as complementary rather than conflicting. Susan Leigh Star, writing with information scholar Karen Ruhleder, asserts that "an infrastructure occurs when the tension between local and global is resolved."[56] For data must not only function within local structures, such as an organization, but also have a broad enough "scope" to serve relevant outside interests. Infrastructures for data are designed to work at a distance, but keeping the local and global relevant to one another requires constant coordination work. Edwards, who writes about global climate models, explains that understanding the relationship between local and global orders is a research problem: "What made it possible to see local forces as elements of a planetary order, and the planetary order as directly relevant to the tiny scale of ordinary, individual human lives?"[57]

At the Arnold Arboretum, data may be shaped by local conditions, yet they serve a combination of needs, near and far. While accessions data help staff to manage the collection on a day-to-day basis, they are also relied on by researchers from institutions around the globe who collaborate with local arboretum staff as well as by tourists who visit from abroad and delight in using data tagged directly on the plants to navigate the otherwise-exotic collection.

Nevertheless, there is no global experience of data, only an expanding variety of local encounters. Data travel widely, but wherever they go, that's where data are. For even when data escape their origins, they are always encountered within other significant local settings.[58]

Local Is Not an Unquestionable Good

The term *local* is a point of pride in advertising. Local ownership, sourcing, and artisanship are all reasons that one should, purportedly, choose to "buy local." Moreover, as recent criticism of globalization gains momentum, local can come to be seen by some as unequivocally positive. But local can also mean exclusionary, narrow, or even oppressive. In recent years, right-wing nationalist movements in the United States have sometimes aligned themselves with localism. Many examples in this book reveal how local conditions can create problematic data or data practices. The DPLA offers opportunities to see how data created to manage collections of cultural history can reify racist social categories that are local to places in the United States. Why is it that curators at the Smithsonian, for instance, hold data on the race of "black" and "African American" artists, but not those of "whites"? This absence paradoxically legitimizes white supremacy by suggesting that whiteness is normal, not in need of special attention, and the default from which all other racial categories must be distinguished.

Returning to the example of the news, we now know that choosing to see only what is local can lead to the emergence of filter bubbles—a phenomenon that online organizer Eli Pariser explains can arise when data-driven algorithms return only the results that we expect.[59] All other perspectives are "filtered" out. Filter bubbles have been blamed in recent years for the extreme polarization of US politics, leading Democrats and Republicans alike to dismiss the other side as uninformed, or even misinformed by what they deem to be "fake news." As we learn to engage with the local conditions of data, we must also become aware of when those conditions need to be protected and when they are more justifiably challenged.

Local Often Means Invisible

A local perspective is not easy to maintain, particularly when interfaces to data ask us to accept those data out of context. Appreciating that all data are local requires effort. Unfortunately, minding and maintaining the context of data is frequently invisible labor, best understood as a kind of care.[60] Feminist scholar and political theorist Joan Tronto, writing alongside Berenice Fisher, defines care as "everything we do to maintain, continue, and repair 'our world' so that we can live in it as well as possible. That world includes our bodies, ourselves, and our environment, all that we seek to interweave in a complex, life sustaining web."[61]

Seeing data work as care is in keeping with Harding's feminist and postcolonial focus on the local. Librarians and archivists explicitly engage in care through data. In chapter 3, I write about conflicts between librarians and internet scholars in the formation of the DPLA. Librarians "deal in privileges," one of the DPLA staff warns, adding, "They keep data away from the public."[62] Another cautions, "Librarians can be overly concerned with the value of their holdings and themselves."[63]

Such comments are disparaging of important care work that is necessary for the creation, operation, and continued maintenance of data in every setting. We must all do more to actively care for our data and any vulnerable subjects that they represent. When such work is derided or undervalued, it perpetuates a long history of degrading care.[64] Seeing data as local necessitates acknowledging care for data as well as those who carry out that work. Especially for those readers who identify as men, as I do, adopting a feminist perspective can be a crucial step toward challenging gendered notions of objectivity applied to data. Treating data subjects abstractly, avoiding intimate relationships with those subjects, and holding them at a comfortable distance does not make one more rigorous but rather less well informed.[65]

CONCLUSION

In practice, accepting that all data are local means engaging with data settings instead of simply data sets. Today, data are too often harnessed as discrete tools to enable analytic work at a distance. The perception of data as immutable yet mobile sustains this view. But we don't have to distance ourselves to make use of data. Indeed, data have been used in many instances—by scientists and curators, among others—as a means of keeping subjects close across long stretches of space and time. Local practices necessitate forming close relationships with not only data but the conditions in which those data are manifest too. Thus, this book is meant to not only reflect on the representational and rhetorical aspects of data from a distance—what values and assumptions they embody—but also engage directly with the knowledge systems that data construct and maintain. What does it mean to take a local stance in data gathering and analysis? How can we get local audiences to care about data? What do models of local practice with data look like? These are some of the questions implicitly posed throughout the book.

In the next chapter and first collections data case, I examine the various roles that place can play, using accessions data from Harvard University's Arnold Arboretum, one of the world's largest collections of trees, vines, and shrubs. Its story illustrates complex ways in which data can be about, in, or from a place as well as how the profile of a place might be understood in terms of its data.

2

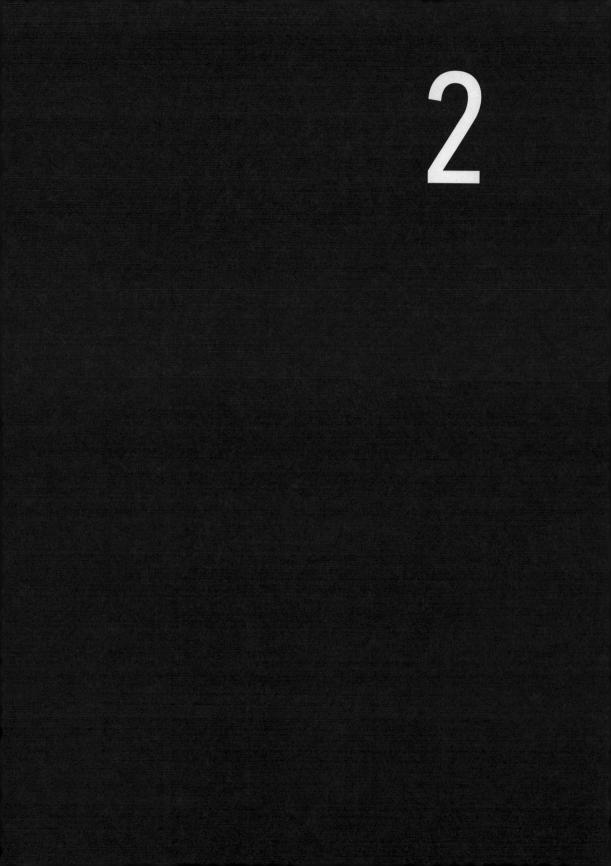

A PLACE FOR PLANT DATA

ACC_NUM, HABIT, HABIT_FULL, NAME_NUM, NAME, ABBREV_NAME, COMMON_NAME_PRIMARY, GENUS, FAMILY, FAMILY_COMMON_NAME_PRIMARY, APG_ORDER, LIN_NUM, ACC_DT, ACC_YR, RECD_HOW, RECD_NOTES,

PROV_TYPE, PROV_TYPE_FULL, PSOURCE_LABEL_ONE_LINE, COLLECTOR, COLL_ID, COLLECTED_WITH, COUNTRY_FULL, SUB_CNT1, SUB_CNT2, SUB_CNT3, LOCALITY, LAT_DEGREE, LAT_MINUTE, LAT_SECOND, LAT_DIR, LONG_DEGREE, LONG_MINUTE, LONG_SECOND, LONG_DIR, ALTITUDE, ALTITUDE_UNIT, DESCRIPTION, COLLECTION_MISC

75–67, T, TREE, 617, AILANTHUS GIRALDII, AIL. GIRALDII, GIRALD, AILANTHUS, AILANTHUS SIMAROUBACEAE, QUASSIA FAMILY, SAPINDALES, 75–67, 25 JAN 1967, 1967, SG, U, UNCERTAIN, MR. THOMAS ELIAS, U. S. NATIONAL ARBORETUM, 3501 N. YORK AVENUE, N.E., WASHINGTON, DC 20002 *VIA*, BOTANICAL GARDEN OF THE UZBEC ACADEMY OF SCIENCE, DSHACHAN ABIDOVI, 232, TASCHKENT, UZBEC SSR, COMMONWEALTH OF INDEPENDENT STATES

***Source*: Arnold Arboretum**[1]

ASSEMBLING A VIEW FROM DATA

On a clear day in 2014, a small group of colleagues and I set out to photograph Bussey Brook Meadow from above, using a digital camera tethered to a weather balloon.[2] The meadow, a small stretch of land located in the Boston neighborhood of Jamaica Plain, is an experiment in urban ecology. It is also part of the Arnold Arboretum: a research collection of plants, vines, and shrubs managed by Harvard University. Bussey Brook was established more than twenty years ago to learn what wild varieties might flourish when left unattended in the contemporary environmental and social conditions of the city.[3] The answer, in part, is in our photographs: a lush composite of native and nonnative species, with the sky-seeking "tree of heaven" (*Ailanthus altissima*) dominating the tree canopy (figure 2.1).[4]

Using low-tech instructions for aerial photography developed by the Public Laboratory for Open Science and Technology (Public Lab for short), we made a harness for a point-and-shoot camera out of an empty soda bottle and secured it to an inexpensive but rugged helium-filled balloon.[5] The camera was set to continuous capture. As our rig ascended, it took more than a thousand photographs, each successive shot framing an expanded view of the meadow below.

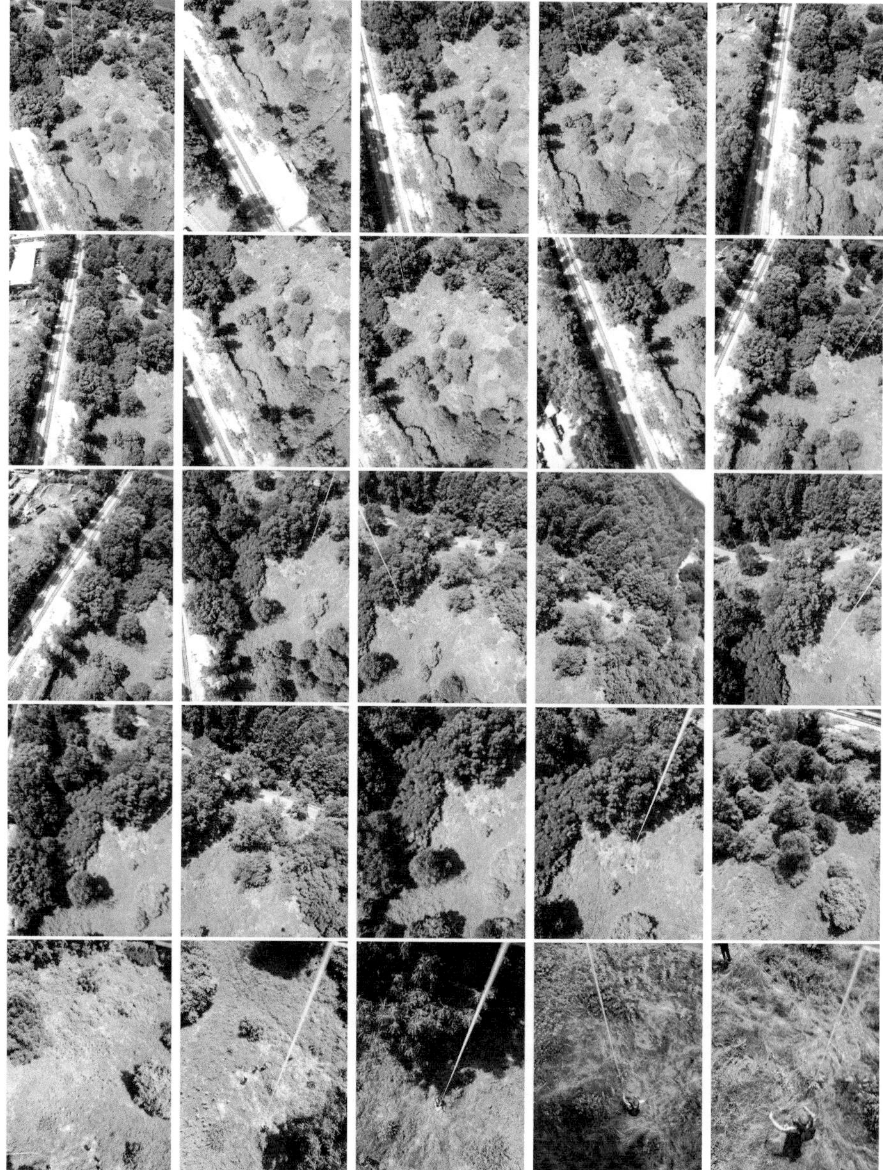

2.1

Selected photos from more than a thousand aerial images of the Arnold Arboretum's Bussey Brook Meadow created with a balloon camera. Image by the author in collaboration with metaLAB.

The first few photographs captured only our small group, huddled near the end of a nylon rope trailing behind the balloon. But in the subsequent images, these tight boundaries expanded, encompassing larger and larger swaths of green at each fraction of a second, until the balloon was at its final height of about five hundred feet. At this height, the frame encompassed many things: the taut string tethering the balloon, our team dotting the ground below, a tangle of plants unidentifiable to all but the most expert observer, and a path weaving between banks of foliage. As the camera pivoted with the wind, its frame shifted again to include the extents of the meadow, adjacent arboretum grounds, and Oak Grove train station, which marked our place at the western edge of Boston.

These images aided arboretum staff in understanding and evaluating their urban ecology experiment. The documentation was necessary because Bussey Brook Meadow is the only part of the arboretum that is not carefully curated and cataloged. As such, our photographs helped to shape the overarching project. But more than a collection of plants to be identified discretely, our images revealed the meadow as a place: our presence, the path, and the proximity to the T stop (Bostonian for "train station") give this feral stand of trees an identity that transcends the sum of its parts.[6]

My interest is not in aerial photography per se but instead in understanding how varied forms of data—of which photographs can be an example—shape places, and reciprocally, how those places shape their data. Beginning with photographs of the arboretum helps us think about what's missing from our understanding of another kind of data: the accessions records of the arboretum's curated collection. We are used to identifying places through images, such as those taken over the meadow. But we cannot say the same for other forms of data. What does it mean to see a place through collections data?

ENCOUNTERS WITH DATA: ABOUT, IN, AND FROM A PLACE

Data have complex attachments to place, which invisibly structure their form and interpretation. This is the second of six principles that frame the book. Place is routinely overlooked as a dimension of situatedness in social studies of data. As I remarked in chapter 1, often but not always, *situated* refers to embodiment or social context. In this chapter, I make use of the arboretum's accessions records to illustrate the manifold relationships between data and place.

Established in 1872 and located on 281 acres, the Arnold Arboretum is equal parts urban laboratory and "zoo for plants" (figure 2.2). It is one of the most comprehensive, well-documented collections of its kind in the world.[7] Its accessions records are one of many genres of data in use at the institution. I have chosen to work with these records rather than other forms of data collected at the arboretum, introduced later in this chapter, because of the evocative ways in which accessions details highlight important place attachments that data can hold.

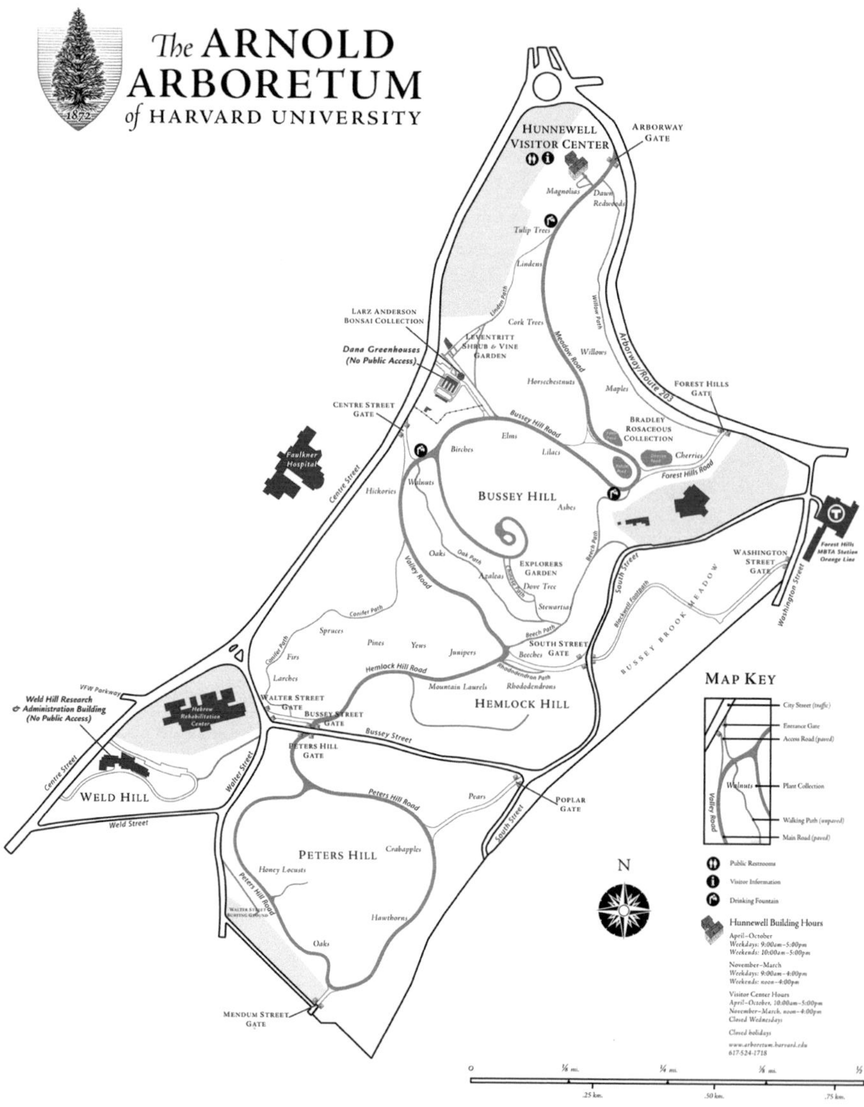

2.2

Map of the Arnold Arboretum. Bussey Brook Meadow is on the right. Courtesy of the Arnold Arboretum.

Hosting around fifteen thousand living plants today, and about seventy thousand over the course of its history, the arboretum is an apt site for investigating data's attachments to place for four distinct reasons. First, place is a key piece of data for the arboretum. Its collections are assembled from sites of scientific and cultural significance around the world. Second, the arboretum has long sought to be a place in which scientists and citizens alike can encounter large collections of data firsthand simply by walking the landscape, and discovering a variety of carefully tagged trees, vines, and shrubs. Third, when understood as a set of conditions for production, place has shaped fluctuations in botanical data over the course of the arboretum's long history. Fourth, when looked at graphically, the arboretum's data can help us see place in new ways, which aren't limited to aspects of geolocation.

In summary, data can be about place, in place, from place, and even generative of place. Learning about these long-standing forms of place attachment can prompt us to challenge settled conceptions about the relationship between data and place in contemporary life. Each of these attachments to place is a different way in which data are subject to local examination. The dimensions of place attachment identified in this chapter along with the means of identifying them suggest a place-based approach to understanding initially unfamiliar data sets in terms of their settings. Let us start by taking a look at how local readings of data related to individual specimens might reveal diverse place attachments.

Place as Data: The Case of *Prunus Sargenti*

> PROV_TYPE, PROV_TYPE_FULL, PSOURCE_LABEL_ONE_LINE, COUNTRY_FULL, SUB_CNT1, SUB_CNT2, SUB_CNT3, LOCALITY, LAT_DEGREE, LAT_MINUTE, LAT_SECOND, LAT_DIR, LONG_ DEGREE, LONG_MINUTE, LONG_SECOND, LONG_DIR, ALTITUDE, ALTITUDE_UNIT

The fields listed above (beginning with PROV_TYPE and ending with ALTITUDE_UNIT) all contribute to the characterization of place in arboretum accessions data. In order to understand the origin of a single specimen using these data, it is necessary to take account of multiple fields and how they might be interrelated. Already mentioned in the introduction, a special cherry tree (*Prunus sargentii*) accessioned to the arboretum on a leap day in 1940 provides an example of this process. The history of the tree is cataloged in a custom digital record system called BG-Base, under the specimen number 130–40. The provenance of the plant (PSOURCE_LABEL_ONE_LINE) is attributed to the institution's founding director, Charles Sprague Sargent, at the address of the arboretum itself: "125 The Arborway, Jamaica Plain, MA." Part of the tree's Latin name, *sargentii*, honors this parentage.[8] Meanwhile, its country of origin (COUNTRY_FULL) is listed as "Japan." Sargent might have acquired the plant during an expedition to Asia. But this would seem to be in conflict with other known conditions—first and foremost, that Sargent died in 1927, thirteen years before the listed accession date. Moreover, for reasons that will be explained later, wild plants from abroad had not been taken in at the arboretum since the mid-1920s.

Sargent could not have transported the plant from Japan to the arboretum on the date of accession. This apparent inconsistency is an artifact of the way that origins are documented at the long-lived institution. Staff at the arboretum know that a few select fields—date, place of origin, and provenance—don't tell the whole story. One has to look to another field, the provenance type (PROV_TYPE) of the cherry tree, to learn that it is a "cultivated plant of known (indirect) wild origin" or "Z" for short. In other words, specimen number 130-40 grew from a cutting taken off another plant collected much earlier. Provenance type is a classification of disputed value, for it is a social distinction versus a biological one. Wild plants and their cuttings are genetically identical. In this case, the "Z" helps to clarify that the cherry tree in question was grown from a cutting of one of Sargent's original specimens—probably number 16760, unearthed from its native Japanese soil in 1892. The apparent difference in the formats of these two specimen numbers is just another indication of how data creation practices can change over the life of an institution. The former (130-40) has a two-digit year attached (the part after the dash) to indicate that the specimen was accessioned in 1940. The latter (16760) was created before that practice was adopted.

This example illustrates some of the complexities of place as presented within the arboretum's data. Data *about* place are not simply contained in a field. This form of place attachment must be understood through a matrix of values, coordinated through local knowledge about the history of data collection practices and how they encode place as a subject.

Place of Data: The Case of *Torreya Grandis*

As the previous instance shows, the arboretum is an aggregated landscape stitched together from plants once residing in other places. Most plants hail from ecological zones similar to that of Boston's, stretching across England, Greece, South Korea, China, and Japan. When encountered at the arboretum, each of these plants stands with its data. A thin plastic card embossed with a subset of accession details hangs from its trunk or branches (figure 2.3). The cards contain fields that are relevant for arboretum staff, researchers, and visitors: scientific name, accession number, plant family, accession date, propagation material (e.g., seed "SD" or scion "SC"), location, common name, and source/collection data. Together the plants and their tags transform the arboretum into a full-scale scientific map organized using the Bentham and Hooker taxonomy—a system that dates to the late nineteenth century. The arboretum landscape is itself a place for encounters with data.

In order to understand this second form of place attachment, let us revisit a tour of the grounds that occurred in late June 2013. During a workshop that I coorganized, a group of visitors were guided by arboretum senior researcher Peter Del Tredici through the Explorer's Garden, an area nestled in a microclimate beneath the summit of Bussey Hill. Del Tredici stopped to comment on his relationship to the living collections. "I've got a lot of direct connection to a lot of these plants. That little plant, *Torreya grandis*, I collected in China in 1989. So a lot of these are like my offspring."[9] Del Tredici explains that he found the seeds of the *Torreya grandis* at a market in China. Fleshy and green, they struck him as unusual examples of edible seeds produced by a conifer. But beyond what is interesting about the plant itself, this quote provides a compelling starting point for understanding what is and is not included in the information landscape of the arboretum.

The acquisition date of the *Torreya grandis* and Del Tredici's association with it are duly noted on the plant's tag. Also pressed into the tag's smooth surface, the term *pinales* registers the plant's status as a conifer. There is no hint of what Del Tredici has referred to as the "oddness" of this ordering.[10] Indeed, several features of the plant's local significance are not included on the tag, which serves mainly to position the *Torreya grandis* within a scientific landscape. Tags do not explain how plants like this one are literally and figuratively torn up by the roots, and then relocated to a new ecological and cultural context. Let's explore a few of the conditions that don't count as data in this context.

Del Tredici is identified on the tag as a "collector," not as "progenitor" or "breeder," as his statement would suggest—this, despite the fact that he is responsible for the reproduction of the plant in the Boston region. The term *collector* speaks of the scientist-and-specimen relationship between Del Tredici and the plant, rather than the more nurturing association between Del Tredici the horticulturalist and the organism he has cultivated. The latter is more in line with his own intimate way of identifying the *Torreya grandis* as his "offspring."

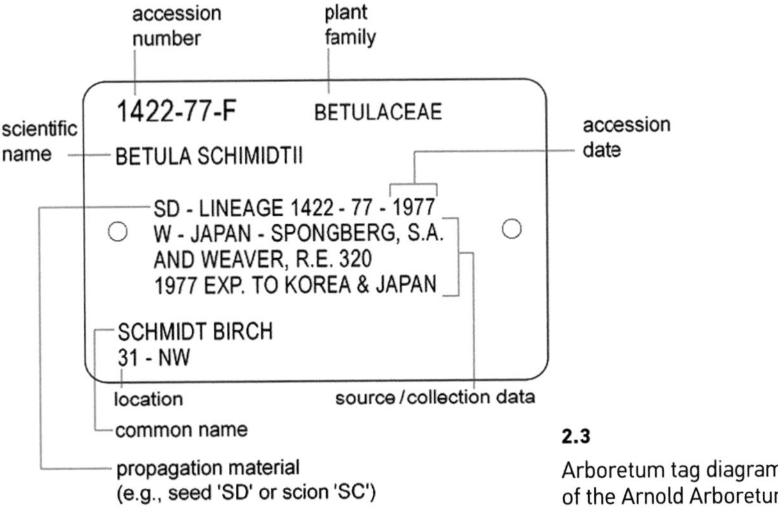

2.3

Arboretum tag diagram. Courtesy of the Arnold Arboretum.

More generally, there are few traces of the fruitful intersections between the living collections and local communities in Boston that surround the arboretum. Don't look to data for connections between "dandelions" (*Taraxacum officinale*) and the elderly Greek women who collect them from the arboretum grounds in the early summer to make *horta vrasta* (boiled greens), or associations between the "tree of heaven" (*Ailanthus altissima*) and the devout Dominicans who discover starlit sites for their Santeria rituals in the groves of Bussey Brook Meadow. Such details, though important to the local meaning of the arboretum's plants, are not part of the way that data on tags interact with the place.

I introduce the example of the *Torreya grandis* to call attention to the placement of data, but also their limits as tools for understanding the places in which they reside. While useful as a means of establishing shared references among the arboretum's staff and its visitors, data do not capture the full lives of arboretum plants.[11] Data categories sit beside yet do not account for all the varied place-based meanings that the plants embody.

Data of Place: The Case of *Tsuga Caroliniana* and *Tsuga Canadensis*

So far I have revealed how place appears in data, and how data appear in place. There is also the important matter of how a place affects data's production. For this last point, let us consider the Carolina and eastern (or Canadian) hemlocks (*Tsuga caroliniana* and *Tsuga canadensis*), which are trees local to the East Coast of the United States. Both have been in rapid decline due to a nonnative insect, the hemlock woolly adelgid (*Adelges tsugae*). In April 1997, an unaccessioned stand of almost two thousand hemlocks on Hemlock Hill—originally brought to the arboretum many years earlier, not as scientific specimens, but as filler plants meant to occupy a bald spot on the landscape created by a destructive hurricane—fell victim to the pest. A note in the accession record for one Carolina hemlock reads, "Plants producing very heavy seed crop, heavily infested with woolly adelgid."[12] Over the winter of 1997–1998, the trees were "labeled, mapped, and qualitatively assessed" to monitor damage caused by the infestation.[13] Although these trees had been residents on the institution's grounds for decades, they were only accessioned into the collection in order for the infestation to be tracked and treated with imidacloprid, a powerful insecticide. The hemlocks were never meant to be an official part of the collection. Regardless, the accession of the blighted hemlocks made 1997–1998 a peak moment of expansion for the arboretum, but only from the perspective of data.

This example demonstrates that even seemingly straightforward fields like "date" can have a complex relationship to place. For each entry in BG-Base, what the accession date means is dependent on local conditions. It might mean when a seed was planted, when a seedling arrived on site, or simply—as in the case of these hemlocks—when an existing plant was annexed to the collection. But beyond the curious and local significance of their accession dates, the hemlocks are interesting because they raise deeper issues about the role that data perform at the arboretum.

Controversy still surrounds the decision to make the stand of hemlocks part of the collection. Del Tredici, who originally argued for their accession, continues to see the trees as invaluable for the study of the infestation process. "It was only by accessioning the plants that we could track their decline over time or the insecticidal treatment."[14] Meanwhile, current director William Friedman, who arrived years after the hemlocks were accessioned, looks on these trees of questionable provenance as inherently undesirable, for they lack essential data about their origins that would make them reliable subjects of scientific study. Why not replace them with trees of actual research significance?

Such disagreements highlight the tensions between competing realities at the arboretum: it is a living place, but also a repository for data. Hence data may be looked on as "just good enough" to support the care of the collection: organizing plants, notes, and relationships among them in a convenient manner.[15] But without reliable data, the emergent form of the collection can disappear altogether, with its contents scattered in an ontological wild.

Coexisting concerns about the necessity of data and their inherent instability over time reinforce a lesson from STS that holds across shifts in technology: data must be part of a knowledge system, or what Paul Edwards calls a "knowledge ecology."[16] The connection to environmental processes is apt. Arboretum scientists, specimens, and information infrastructures are all necessary to generate, verify, and sustain what the place knows. It is the encompassing place—of which data are only a part, along with the people and plants—that holds knowledge about the arboretum hemlocks, their deadly infestation, and its implications for similar trees across the Northeast. At the arboretum,

2.4
A hemlock tree at the arboretum. Image by the author.

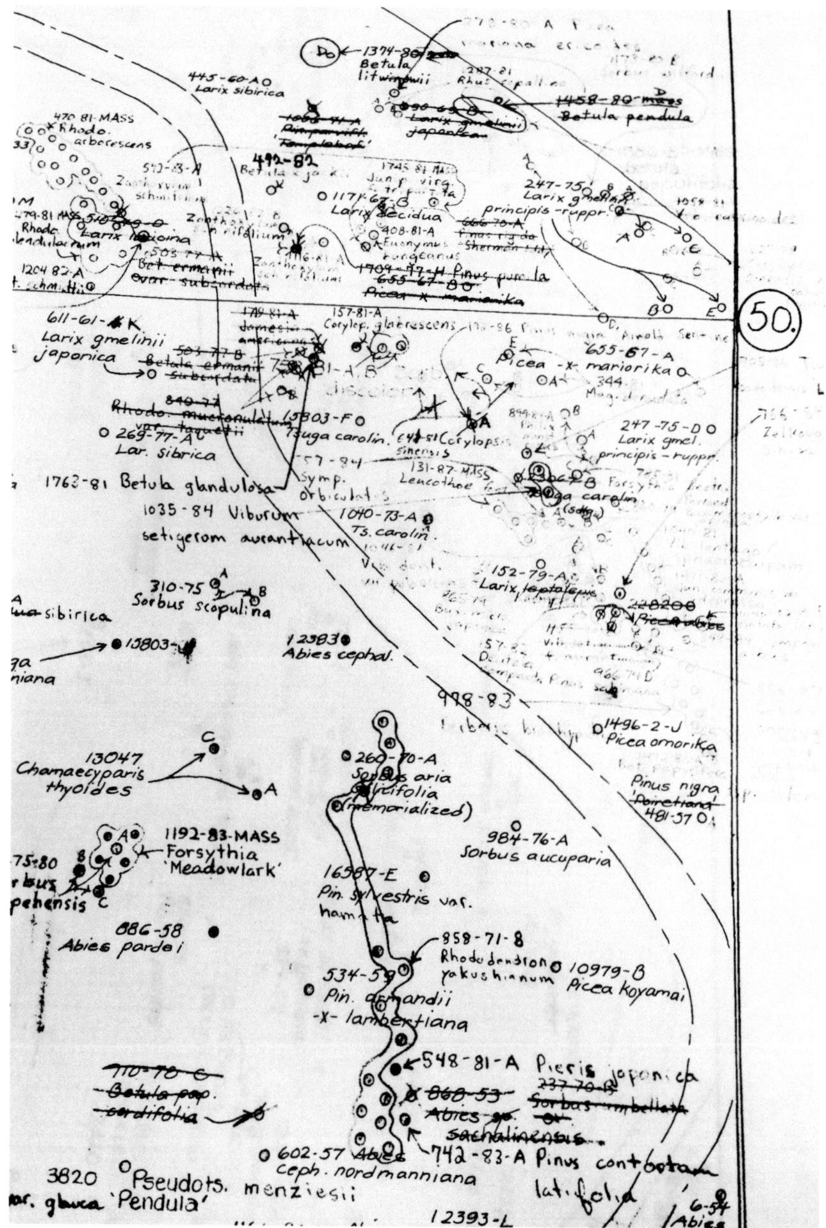

2.5

Early map of the arboretum.
Image by the author.

the knowledge ecology is more than a metaphor. Data are necessary components of the functioning ecology created and maintained there. The story of the hemlock trees (figure 2.4) illustrates how becoming data can be a prerequisite for receiving sustaining care. Thus data can transcend their roles as representations by directly supporting the places that they describe.

READING DATA IN PLACE

In order to create these local readings of accessions data, I have relied on a prolonged engagement with the Arnold Arboretum. During the period from 2012 to 2014, I lived and worked in close proximity to the institution in Jamaica Plain. I conducted interviews with researchers, administrators, and technologists, and searched through archival sources at their library. More important, though, I was a participant observer in both formal and informal engagements, including a course on landscape architecture taught by Del Tredici, the series of outings to photograph Bussey Brook Meadow, and the multi-day workshop that I coorganized to bring together arboretum staff with scholars of science and technology. Over the course of the final year of this engagement, I worked with the staff to develop reading techniques appropriate for looking at their data.

Seeing data as texts accessible to traditions of hermeneutic inquiry means reading them within an interpretative context. It would be difficult to understand these records without considering their historical attachments to the arboretum as a place. Indeed, its accessions records have a long history of development and use. For one thing, they were not always recognizable as data. The arboretum has weathered many successive regimes of documentation (figures 2.5–2.7).

Each organism at the institution has germinated within a social and technological setting, its care and curation managed through the instruments and information structures deployed during its lifetime. These place-based practices along with the documents that they produce register what is valued about individual organisms and, in turn, how those values change over time.

Today, plants collected from around the world and across time are held together by BG-Base. Each entry in the arboretum's data set includes an accession number, an extensive list of scientific, common, and abbreviated names, redundant ways of identifying the time of accession, the form and mechanism of reception, individuals associated with the plant, various descriptions of the place that the accession hails from, its condition in the wild, and an additional catchall category. A list of fields used by the arboretum includes:

> ACC_NUM, HABIT, HABIT_FULL, NAME_NUM, NAME, ABBREV_NAME, COMMON_NAME_PRIMARY, GENUS, FAMILY, FAMILY_COMMON_NAME_PRIMARY, APG_ORDER, LIN_NUM, ACC_DT, ACC_YR, RECD_HOW, RECD_NOTES, PROV_TYPE, PROV_TYPE_FULL, PSOURCE_LABEL_ONE_LINE, COLLECTOR, COLL_ID, COLLECTED_WITH, COUNTRY_FULL, SUB_CNT1,

SUB_CNT2, SUB_CNT3, LOCALITY, LAT_DEGREE, LAT_MINUTE, LAT_SECOND, LAT_DIR, LONG_DEGREE, LONG_MINUTE, LONG_SECOND, LONG_DIR, ALTITUDE, ALTITUDE_UNIT, DESCRIPTION, COLLECTION_MISC

If encountered within a library, museum, or archive, many of these fields would be considered metadata: the information necessary to catalog a book or other object, such as details of their contents, context, quality, structure, and accessibility. At the arboretum, this locally defined selection of fields is known simply as "accessions data." But they are shaped by many of the same forces that affect metadata.[17] For example, each accession record exists as part of a local constellation of information, including the details of the associated plant's phenology, genetic characteristics, transpiration rate, and growth habit. Even the specimen itself is a kind of data.[18] This entire "data assemblage" is necessary to make plants real as well as present in the contemporary ecological, scientific, and public life of the arboretum.[19]

As mentioned above, documentation practices at the arboretum long predate contemporary notions of data. Today, records are available in multiple formats simultaneously: on maps (figure 2.5), in ledgers (figure 2.6), on index cards (figure 2.7), and only recently, in digital form. It wasn't until summer 1985 that the arboretum started converting its accessions data from index cards crowded in a vertical file to digital data stored in BG-Base. These digitized data afford new opportunities for access and

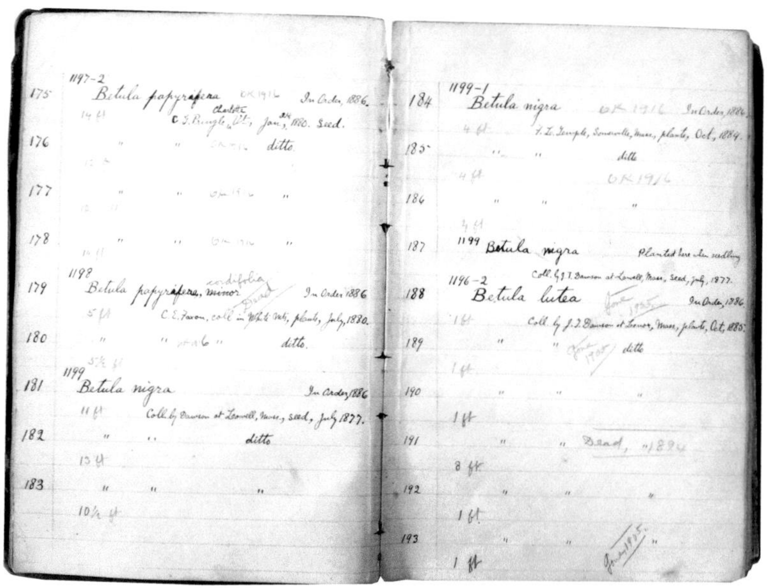

2.6

Early ledger containing accessions information. Image by the author.

analysis. Even so, some staff members continue to use older formats exclusively for they do not yet trust the process of digitization or interpretations of outsiders with newfound access to their data.

Regardless of the format, what counts as data at the arboretum is a matter of context. As Del Tredici explains, "The data, in and of itself, is only valuable [for] somebody who understands its significance." To further his point—one that I have tried to echo throughout the book—Del Tredici likens the "raw data" to seeds. When a seed won't germinate, there are innumerable possible reasons. "Unless you know how to interpret the behavior of the seed, it is just nondata."[20]

VISUALIZING PLACE

Through my local readings of the Arnold Arboretum's accessions records, I have sought to reveal numerous ways in which data can be entangled with place: when place is a kind of data, when place is the site of encounters with data, and when place is the site of data's production. Each of these place attachments can be exposed through local readings of accessions data for particular plants: *Prunus sargentii*, *Torreya grandis*, and *Tsuga caroliniana*. These case studies were triggered by discrete technical problems or controversies that I happened upon. As such, they provide a kind of event-based reading of data. Yet looking at the accessions of the arboretum altogether through visualization techniques can reveal alternative conceptions of place.[21]

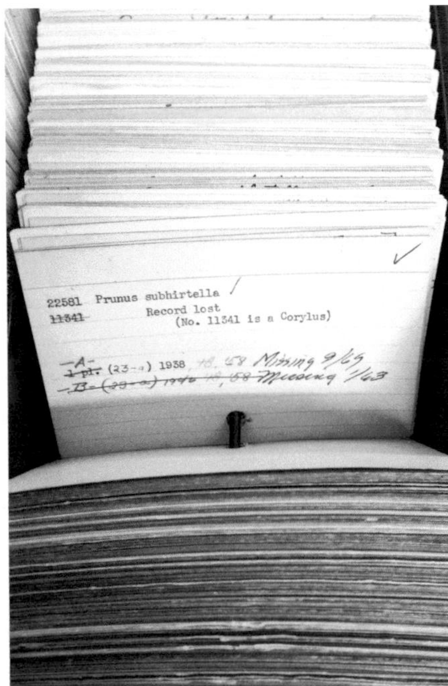

2.7

Card catalog containing accessions data before it was digitized. Image by the author.

I use the term *visualization* to describe the experience of looking at the whole arboretum through the data of its parts. Rather than being a god's-eye view, characterized by theorist of science and technology Donna Haraway as one that seems to come "from nowhere, from simplicity," these visualizations offer situated but wide-ranging perspectives: views of data as opposed to views through data.[22] Creating this kind of visualization requires a critical sensibility toward data, including attention to what might be occluded and what other vantage points are possible. This approach complements prior work in geography on the critical studies of landscape representation as well as the development of critical practices in mapping.[23]

My visualizations are meant to function more like panoramas than maps. Although it is not uncommon to hear the term *panorama* used today to describe graphical displays of data, few acknowledge that unlike maps, panoramas are situated ways of seeing places. As far back as the eighteenth century, the term was used to describe pictorial representations of landscapes as seen by an observer positioned at a single strategic point. Moreover, panoramas have long been understood as mediated. Like visualizations, they are enacted through technological means. For instance, historian Wolfgang Schivelbusch uses the term *panoramic* to evoke the once-unfamiliar view across an expansive landscape afforded by the speed of the passenger train.[24] Just as the rapid pace of the passenger locomotive offered new vistas across broad stretches of space, the visualizations included here reveal perspectives at previously incomprehensible scales. Visualizations aren't narrowly defined technical tools; they generate alternative experiences of data and the places that they depict.[25]

In the visualization presented below (figure 2.8), the arboretum is portrayed as an aggregate, pattern, and system in flux. Here, data are enlisted to construct a new sense of place. Because of their scale and heterogeneity, large digitized data sets offer opportunities for experiences of place that—like Schivelbusch's locomotive panorama—are different from anything seen before.

In the next section, I introduce a series of experimental visualizations. None of these are neutral or inevitable. Rather, they help us reimagine the arboretum as a place with origins, structures, and dynamics that are not merely geographic.

Place as History

Figure 2.8 portrays the arboretum as an aggregate place developed over time. Beginning in 1872 and ending in 2012 (when this set of records was made available for use), the visualization portrays a temporal graph of plant specimens. The image is a kind of timeline, structured by yearly accessions, much like trees record environmental changes in their annual growth rings. Months and days index accumulated plants, each denoted by a dot. This two-dimensional view can be enhanced by a series of section cuts through daily accessions (figure 2.9). In the original interactive version of this visualization, the section cut, which portrays the number of accessions on each day of the selected year, can be produced for any year along the timeline.[26]

Such visualizations can be used to call attention to variations in the data by linking them to color, size, and other visual cues. For instance, figure 2.8 displays changes in provenance type (a category mentioned earlier) across the history of the collection. Here a green dot represents a plant collected in the wild, a yellow dot signifies a cutting from a wild plant, a black dot indicates a cultivated plant, and a gray dot stands in for a plant from an unknown origin (far more common than one might expect). Fluctuations across these provenance-related colors illustrate shifts in the makeup of the arboretum from collections of scientific importance (mostly collected from the wild) to selections in the service of horticulture (mostly from other cultivated collections).

The distribution of green, yellow, black, and gray dots faintly demarcates three eras of collecting identified by curator of living collections Michael Dosmann.[27] In the late nineteenth and early twentieth centuries, Sargent engaged in a global project of scientific fieldwork to collect distantly related species from around the world as evidence to support Charles Darwin's theory of evolution. In the 1920s, however, the US Department of Agriculture discovered that the arboretum was inadvertently collecting invasive bugs along with its imported plants and took action to stop it. Sargent lost the ensuing legal battle, and wild collecting decreased substantially thereafter.

The middle years of the twentieth century are sometimes referenced by staff as the Wyman era, after a prominent horticulturist. During this time, the arboretum halted its foreign expeditions and relocated its scientific research to Harvard's Cambridge campus. The next phase of research centered on the herbarium, a much larger collection made up entirely of dried plants (figure 2.10). Dosmann explains that the expansive grounds in Jamaica Plain became a "showcase garden," a place to display the horticultural trends of the day. In this period, says Dosmann, "if you did want to go and collect anything, you went to a nursery."[28]

It wasn't until the early 1970s, during a reevaluation of the mission of the collection, associated with its centennial, that the arboretum reinitiated its expedition work abroad. The renewal of overseas fieldwork expanded relationships with institutions in Asia, and later, studies on emergent and imperative questions around global climate change. My provenance visualization registers some aspects of these long-term temporal shifts, highlighting in particular the relationship between the two defining arms of the arboretum, scholarship and horticulture, and the ways in which their relationship changed over time.

This visually oriented reading of the arboretum as data is unlike a photograph or geographic map of the place. It highlights a landscape shaped over time by otherwise-invisible ecological, organizational, and even political forces. This particular use of the method is but one way of reading. The data support many alternative portrayals of place.

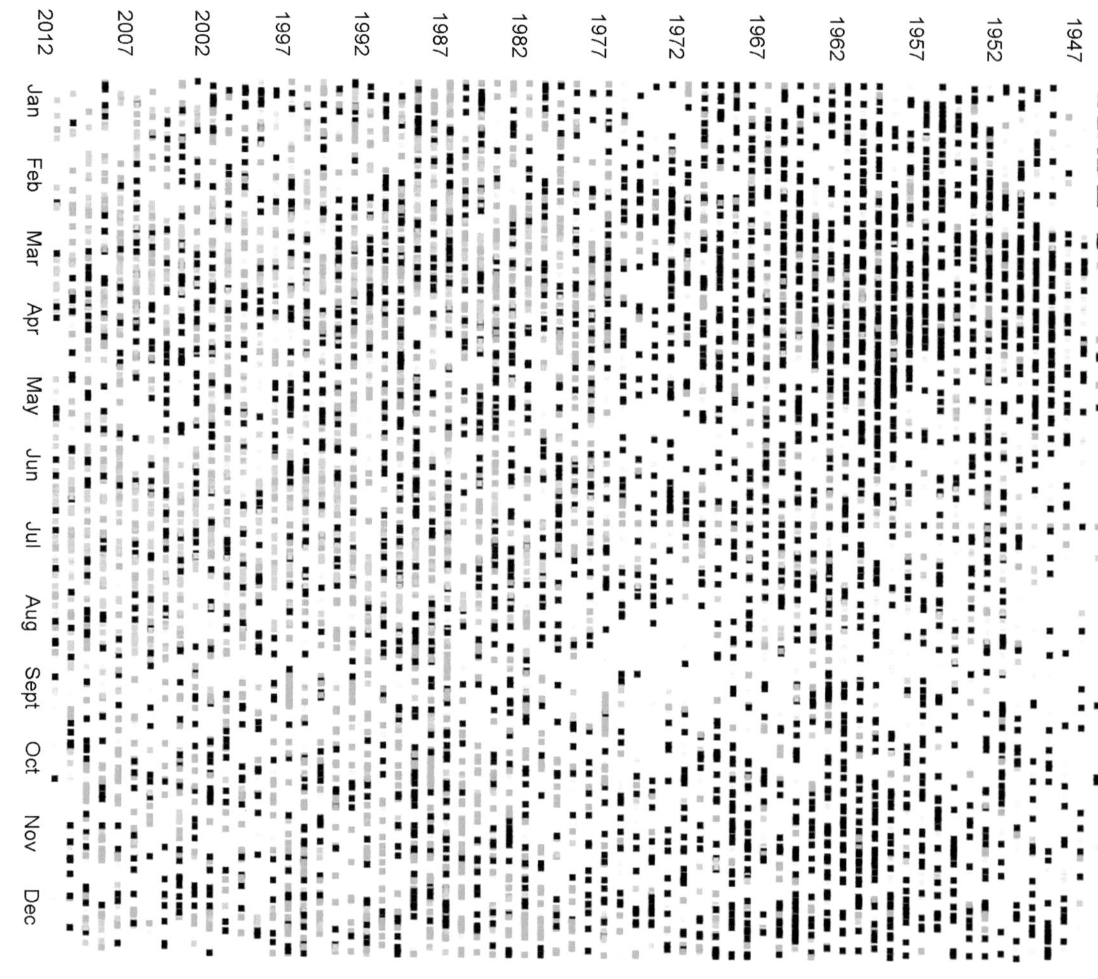

2.8

Linear timeline of arboretum accessions. Plants without accession dates, of which there are 1,190, are not included here. Image by the author and Krystelle Denis.

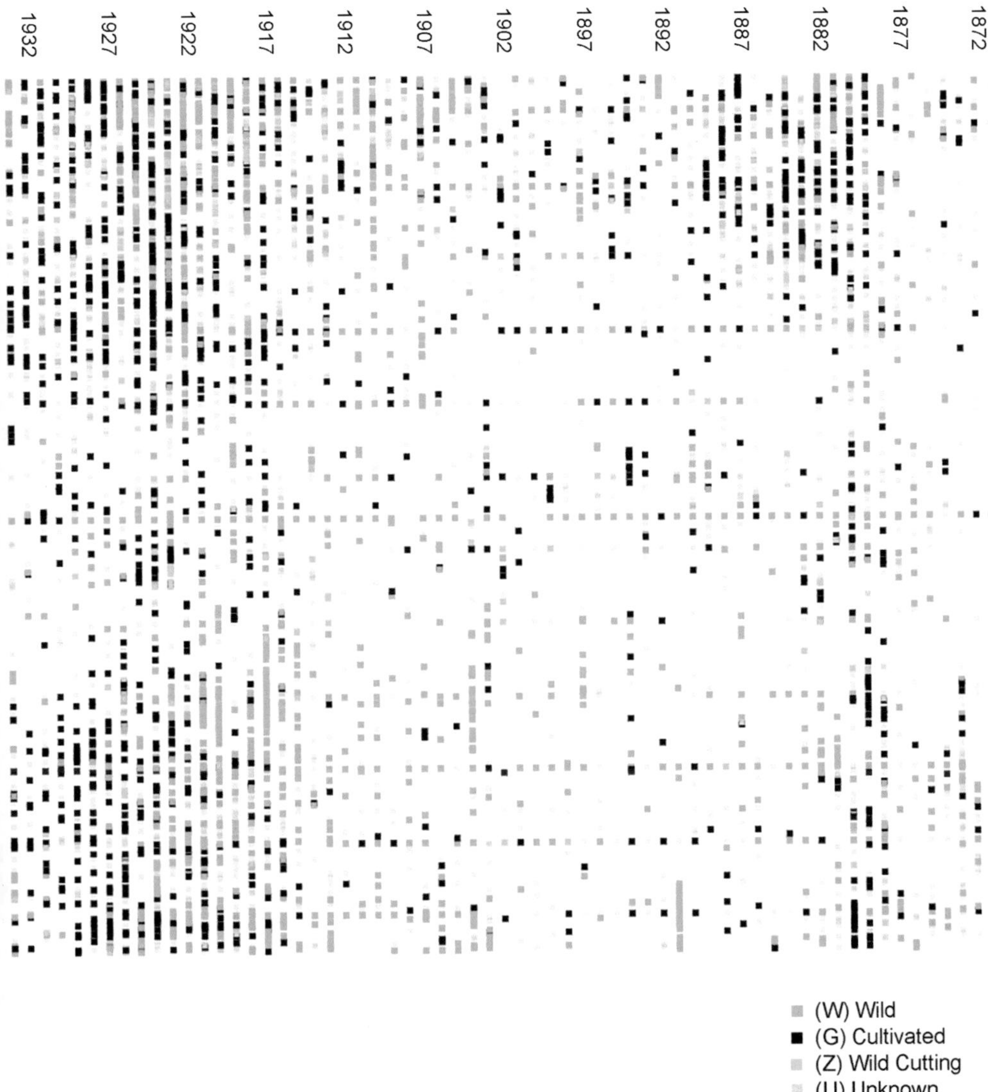

2.9 (following pages)

Section cut through linear timeline of arboretum accessions. Image by the author and Krystelle Denis.

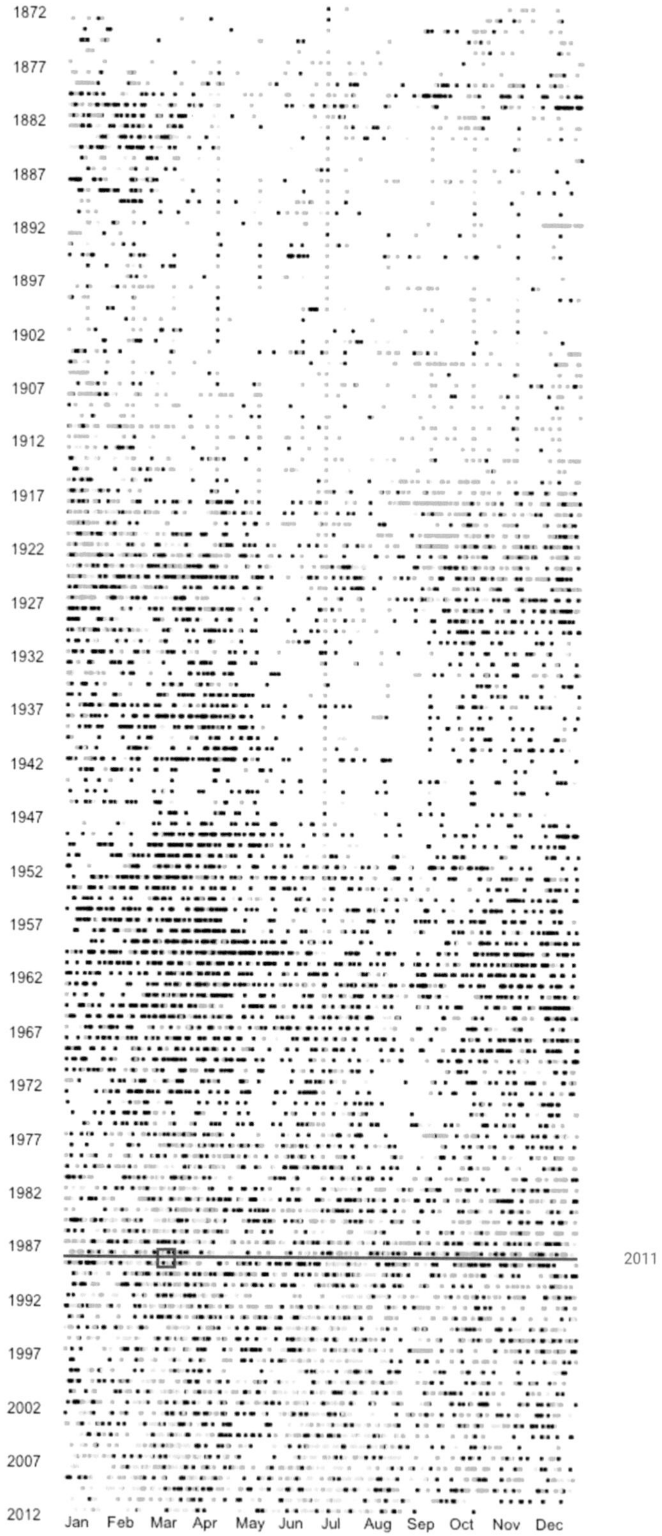

1872
1877
1882
1887
1892
1897
1902
1907
1912
1917
1922
1927
1932
1937
1942
1947
1952
1957
1962
1967
1972
1977
1982
1987 2011
1992
1997
2002
2007
2012

Jan Feb Mar Apr May Jun Jul Aug Sep Oct Nov Dec

8 ACCESSIONS ON 8-MAR-11

44-2011: **MAGNOLIA X SOULANGEANA,** SHRUB/TREE, MAGNOLIA FAMILY, U. S. NATIONAL ARBORETUM, 3501 NEW YORK AVENUE, WASHINGTON, DC 20002, COLLECTED BY: UNKNOWN

45-2011: **SYRINGA PEKINENSIS 'MORTON', CULTIVAR OF LILAC,** TREE, OLIVE FAMILY, KNIGHT HOLLOW NURSERY, INC., 7911 FORSYTHIA COURT, MIDDLETON, WI 53562, COLLECTED BY: UNKNOWN

46-2011: **CERCIS CANADENSIS "SPORT",** BEAN FAMILY, ARNOLD ARBORETUM, 125 THE ARBORWAY, JAMAICA PLAIN, MA 02130, COLLECTED BY: UNKNOWN

47-2011: **CERCIS CANADENSIS, EASTERN REDBUD,** TREE, BEAN FAMILY, MR. MARK KRAUTMANN, HERITAGE SEEDLINGS NURSERY, 4194 71ST AVENUE SE, SALEM, OR 97301-9242, COLLECTED BY: UNKNOWN

48-2011: **MALUS FUSCA, OREGON CRABAPPLE,** SHRUB/TREE, ROSE FAMILY, UNIVERSITY OF BRITISH COLUMBIA BOTANICAL GARDEN, 6501 NORTHWEST MARINE DRIVE, VANCOUVER, BRITISH COLUMBIA V6T 1W5, CANADA, COLLECTED BY: MACPHAIL, J.

49-2011: **MALUS SP.,** SHRUB/TREE, ROSE FAMILY, HORTICULTURE DEPT., MICHIGAN STATE UNIVERSITY, EAST LANSING, MI, COLLECTED BY: BARON, M.

77-2011: **MALUS X ROBUSTA 'ERECTA', CULTIVAR OF MALUS,** SHRUB/TREE, ROSE FAMILY, PROF. CHARLES SPRAGUE SARGENT, DIRECTOR, ARNOLD ARBORETUM, 125 THE ARBORWAY, JAMAICA PLAIN, MA, COLLECTED BY: SARGENT, C. S.

6857: **SYRINGA KOMAROWII SSP. REFLEXA,** SHRUB, OLIVE FAMILY, DR. ERNEST HENRY WILSON, KEEPER, ARNOLD ARBORETUM, THE ARBORWAY, JAMAICA PLAIN, MA, COLLECTED BY: E. H. WILSON

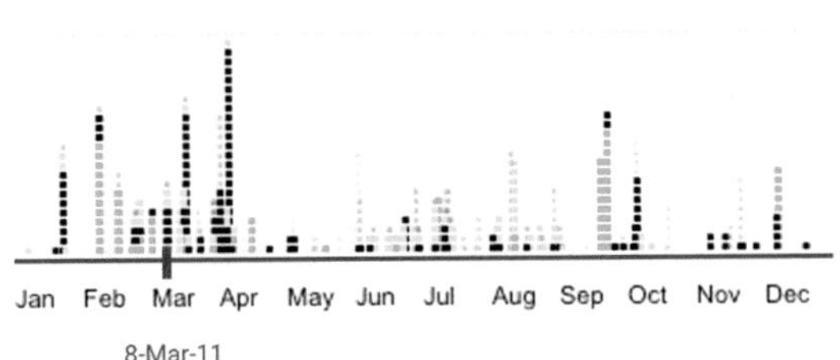

2.10

Torreya grandis herbarium specimen. Courtesy of Harvard University.

Alternative Histories

A radial version of the same timeline pushes the metaphor to embedded arboreal processes—the forming of rings in a tree (figure 2.11). But more important, new patterns are illuminated by the density gradient from center to periphery. The three eras of collecting become more prominent as the sparse accessions in early years are compressed into a smaller space. Subtle lines of accessions running through significant dates of the year are accentuated. They appear as concentrated rays within the circular geometry. The radial organization also suggests an entirely different kind of temporality: one that has an origin at some fixed point and then expands indefinitely into the future.

This radial image can be read against the linear one, which presents time as being infinite in two directions. The accessions depicted in linear form seem sparse in comparison. In the linear version, one can more clearly see increased collecting over the years, albeit with a narrowing in the 1940s. Moreover, practices seem to change dramatically across seasons in the second half of the twentieth century, transitioning from accessioning only in winter to year-round. Other patterns are less visible in the linear timeline. The dispersion of accessions in the early years makes it more difficult to note the intensity of wild collecting during the period of exploration and its symmetry with the period after the 1970s, when the arboretum began to collect externally again. At a more detailed level, a substantial gap in collecting on Christmas day appears clearly in the radial version, but disappears into the fringe of the linear image. This gap could be made more prominent by simply reordering the arrangement of months, but what other patterns would be shifted out of view?

Both the radial and linear versions obscure the exact number of accessions per day. A three-dimensional approach as demonstrated in figure 2.12 can help to make those more evident. Rather than being arranged solely by date, the 3-D image highlights every accessioned plant at the arboretum and exposes the rate of accumulation along a new z-axis. The resulting form is a cone. Moments of rapid growth in the collection appear as narrow sections, whereas periods of slower development flatten it out. While evocative in its shape, the 3-D visualization is more difficult to read. In fact, most of the patterns exposed by other visualizations are compromised in 3-D. Graphics overlap from opposite sides of the cone, the circumference of the yearly rings is visibly narrowed, and daily accessions are difficult to align by month and year.

The above examples of visualization are both interpretative and speculative. They present the arboretum as multiple. Each version of the place offers its own experience of the substantial collections brought together over a long history.

Histories Out of Place

While the visualizations suggest different ways of making sense of the arboretum as a whole, they can also reveal telling details. Indeed, we can learn more about the kind of place that the arboretum is by inspecting components of the visualizations close up.

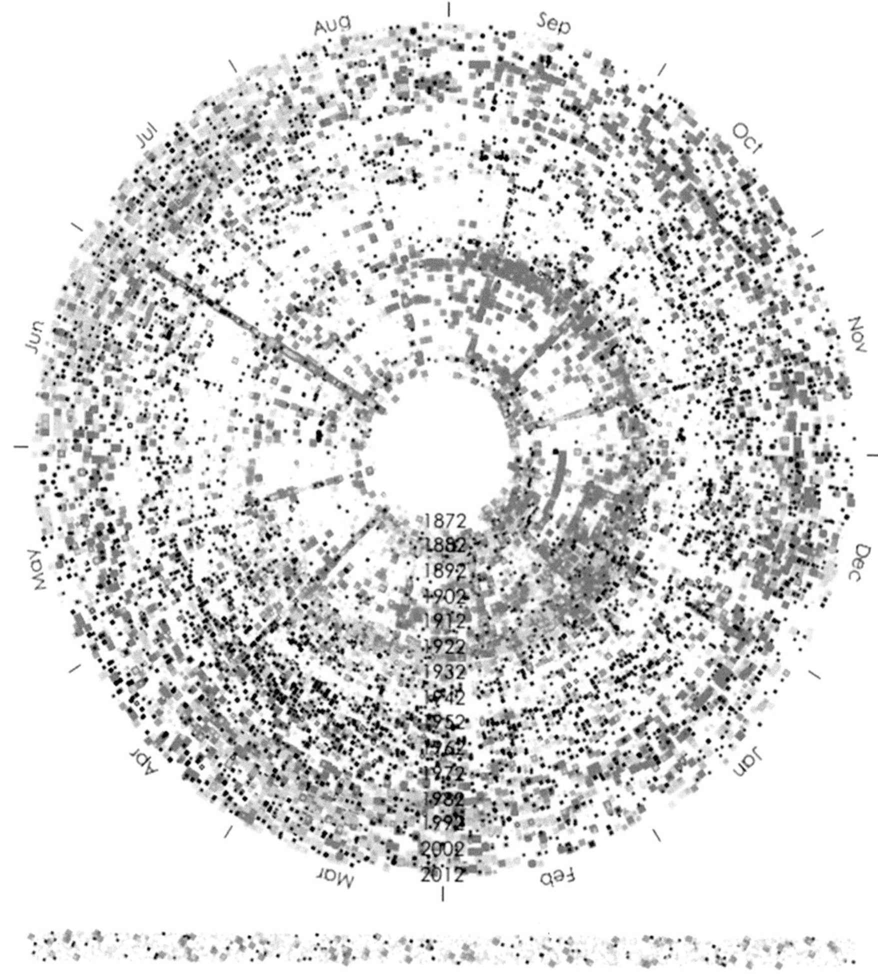

2.11

Radial timeline of accessions to the Arnold Arboretum. The dots on the bottom edge represent plants with no accession date. Image by the author.

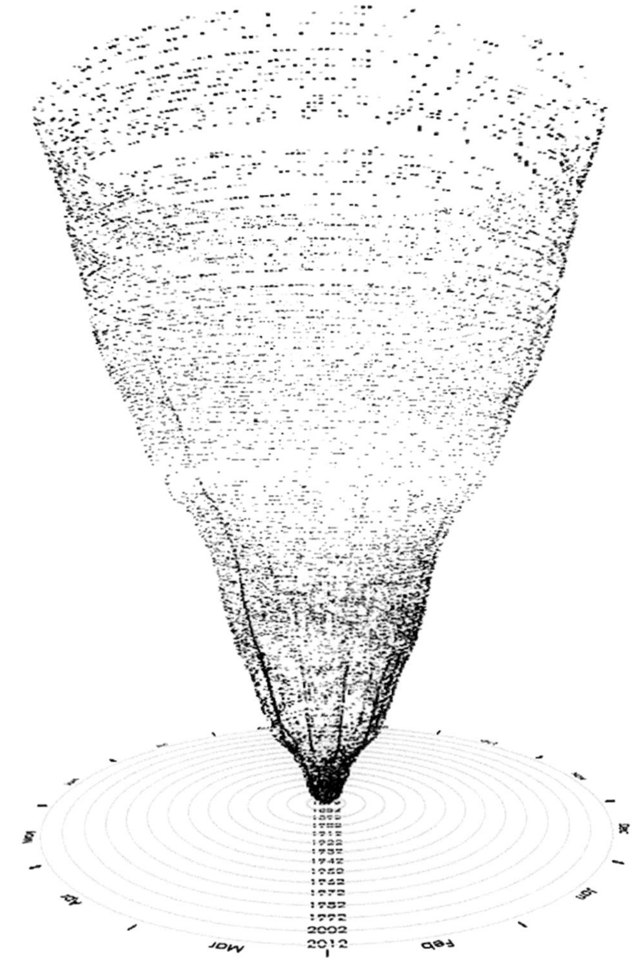

2.12

Three-dimensional timeline of arboretum accessions. Image by the author.

In particular, it is useful to pay attention to apparent anomalies or glitches in the images. I call these "data artifacts."[29] In most work with data visualization, such irregularities are cleaned up. Like various kinds of data dirt, they appear to be simply out of place.[30] But data artifacts speak to the human history of their accumulation.

Consider, for instance, the rays of clustered accessions so prominent in the radial version of the timeline. A literal reading of these rays suggests that accessions arrived en mass on certain days, especially on the fifteenth of every month, the first of the year, and the first of July. Del Tredici, though, suggests that the rays are most likely technological artifacts. "If something came in (during) August 1942, I think BG-Base would output that [by] default as August 15."[31] Without a precisely recorded day of accession, BG-Base places accessions squarely in the middle of the month. The pattern is similar at the scale of the year; accessions appear unusually heavy on July 1, the beginning of the arboretum's fiscal calendar.

Deriving from various processes, such artifacts are often entangled with the contingencies of a place. Those mentioned above might be thought of as production artifacts, resulting from the technical conditions of data creation. Meanwhile, disciplinary artifacts might betray specialized ordering systems, and vernacular artifacts might be the result of dialects or local language uses. These various kinds of artifacts can be extraordinarily subtle and difficult to tease out, but visualization is an adept tool for bringing such conditions to the surface.

Data artifacts register not only local changes in technology, personnel, and organization but also broader cultural rhythms and events. Look closely and you can spot World War II as well as Christmas (mentioned previously) as gaps between denser periods of accessions. The first is manifest as a bald swath in the mid-1940s. The second is particularly noticeable in the radial timeline as a wedge of space radiating down the axis associated with December 25. Accessions from specific regions are affected by international relations too. Del Tredici recounts that "when [Richard] Nixon went to China, I started to get small little exchanges of seed packets and things like that."[32] Through data artifacts, we can see more than a collection of plants. Kyle Port, the arboretum's plant records manager, explains that artifacts betray the "personalities" behind the data.[33] Together, these personalities contribute to the aggregate sense of place generated through visualizations.

A Composite Place

One final visualization offers a view of the arboretum as a collection of places. In figure 2.13, the arboretum is presented as a set of locations extracted from the data. These are not the locations where plants were collected. Rather, they are the addresses of individual collectors. Mapping these data, done here in polar coordinates, results in an image of the arboretum's social network, with each dot representing the home or work address of a collector. The gray-dotted circles call attention to the areas with the largest

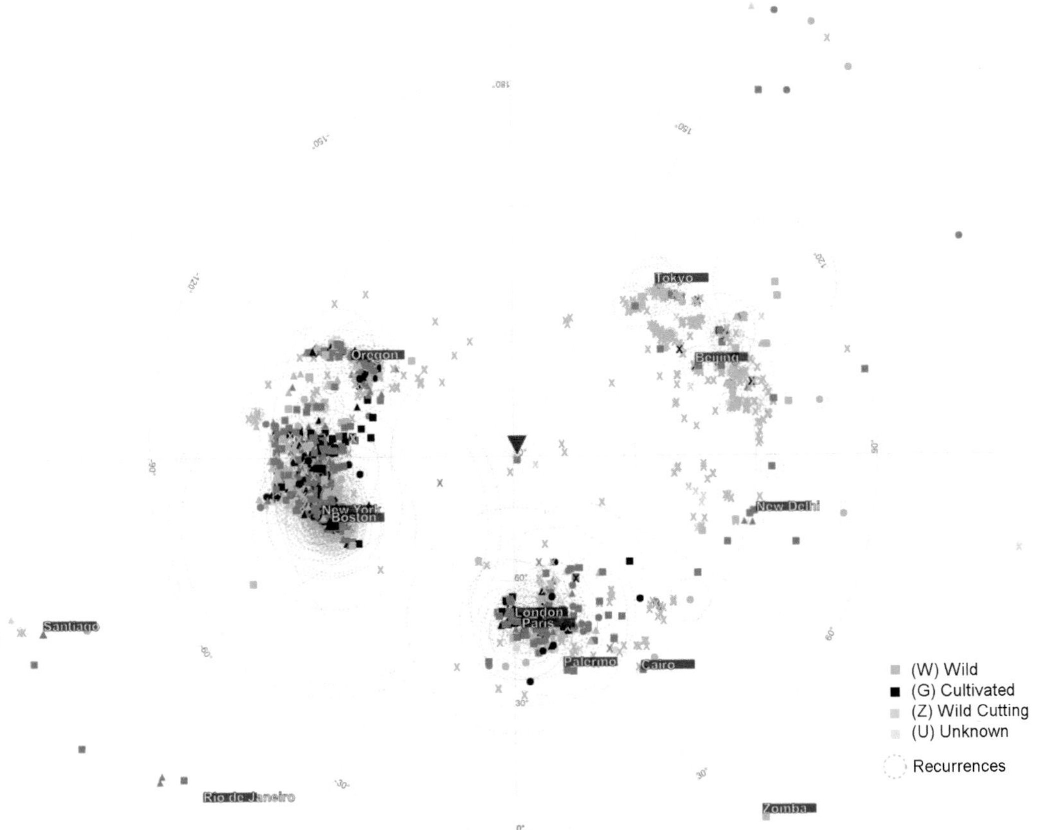

2.13

Map of the addresses of arboretum collectors. Image by the author and Krystelle Denis.

number of collectors, such as the arboretum itself. This way of visualizing the data suggests that the arboretum encompasses an extended landscape of collecting activity.[34] As in some of the previous visualizations, the colors represent provenance type, revealing the professional addresses of the collectors of wild and cultivated plants.

RETHINKING THE PLACE OF DATA

Today, popular media depict data as increasingly commonplace: ubiquitous tools for government, science, and business management.[35] Although there is a long history of scholarship on the *place* of information within discourses on cyberspace, cities, networking, interaction, and development, academic and popular discussions of data frequently downplay the significance of place.[36] Sometimes data and place are treated as incompatible concepts. Geographers Craig Dalton and Jim Thatcher argue that new practices with big data distract from attention to place. "Relying solely on 'big data' methods," they write, "can obscure concepts of place and place-making because places are necessarily situated and partial."[37] What if we learned to see data as situated and partial because of their place attachments?

Although place has been an important topic of interest in the social sciences, my readings of data are also influenced by cultural studies, particularly in the environmental humanities.[38] Lawrence Buell, a leading voice for ecocriticism, expounds on the multiple dimensions of place attachment in texts, including temporal and imagined conceptions of place.[39] My development of the notion of place attachment for data builds on these important precedents, yet it is grounded in readings of data manifest at the Arnold Arboretum.

Here I use *place* to mean an institutionally defined framework with social, technological, and spatial dimensions, in which data are created, displayed, and/or managed, and that reciprocally, is shaped by those practices. Indeed, data are not simply site-specific tools; they have the power to shape place. In common parlance, the term *data* can be used to mean secondary, digital representations of objects that hold scientific and cultural import. But data can also create an ontological "looping effect" whereby they help to shape the practices and institutions that create them.[40]

Do place attachments still hold at the scale of big data? Accessions data at the Arnold Arboretum certainly don't conform to present-day definitions of *big* as high magnitude in a variety of dimensions.[41] Instead, researchers and other staff at the arboretum demonstrate the kind of close relationships with data that contemporary big data approaches were meant to replace, since practices like those carried out every day by Arnold Arboretum botanists require time and proximity that are too often dismissed as expensive and unnecessary.

Having said that, if we consider big data as an epistemological and performative shift in ways of doing research, with a long history involving data sets that were previously unmanageable, we might say that the Arnold Arboretum has been edging toward big data for over a century. In the nineteenth and early twentieth centuries, arboretums—like libraries, museums, and zoos—held the big data of their day. Institutions like the Arnold Arboretum prefigured big data by drawing together representative specimens from far and wide. The most ambitious of these institutions sought to establish themselves as comprehensive models of the world.[42] As with contemporary holders of big data, such institutions of collecting continually outstripped strategies for managing all the records necessary to organize, preserve, and study their subjects. The arboretum's many successive eras of data collection illustrate, better than most, a variety of place attachments in data, regardless of their magnitude.

CONCLUSION

We should learn to see data as cultural forms that are situated socially and technologically, but also in place. Although data are reliably transferred across global communication networks everywhere, they remain marked by local artifacts: traces of the conditions and values that are particular to their origins. Accepting this claim necessitates a significant shift in our expectations of digital data given that the digital was invented to be independent of any substrate.[43] In fact, all data—not just those created at arboretums and other sites for documenting nature—can be read distinctively through their attachments to place.

Each of the place attachments explored in this chapter suggest a different way in which data can be local: by being about, in, from, or even generative of place. Taken together, these four ways of probing data offer a model for how to read data from a local perspective. But this should not be mistaken as a formula for engaging with data anywhere. My methods were developed in situ, with the particular place attachments of the arboretum at hand. Similarly, I encourage readers to challenge these and other settled conceptions of the relationship between data and place by making place a part of their data analysis as well as data presentation.

Exploring the possible relationships between data and place can help us question the wisdom of centralized models of data management. As this chapter has shown, thinking about data as mobile, immutable, and generally detached from place can obscure important ways in which data practices rely on local knowledge as well as experience for meaningful interpretation and responsible use. When taken out of place, data can come to be seen as either the view from nowhere or nothing more than data dirt.

In the next chapter, I explain what happens when data from many places are brought together in the form of data infrastructures. Using the example of the DPLA, a composite collection of digitized media from libraries, museums, and archives across the United States, I illustrate some of the ways in which even displaced and agglomerated data retain traces of their origins, embedded in classifications, schemata, constraints, errors, absences, and rituals that resist simple translation or normalization.

3

COLLECTING INFRASTRUCTURES

{"DATAPROVIDER": "SMITHSONIAN INSTITUTION LIBRARIES, " "ADMIN": {"VALIDATION_MESSAGE": "'RIGHTS' IS A REQUIRED PROPERTY, " "VALID_AFTER_ENRICH": FALSE}, "@ID": "HTTP://DP.LA/API/ITEMS/AF994D5BDBE589A9367E79980BFD7470, " "_REV": "8-138F4B0EE3C24865699469D2A900AAD9, " "OBJECT": "HTTP://LIBRARY.SI.EDU/SITES/DEFAULT/FILES/STYLES/BOOK_COVER_SMALL/PUBLIC/BOOKS/COVERS/CATALOGUEOFGAMES00MILT_COVER.JPG?ITOK=2IK8LJP8, " "AGGREGATEDCHO": "#SOURCERESOURCE, " "PROVIDER": {"@ID": "HTTP://DP.LA/API/CONTRIBUTOR/SMITHSONIAN, " "NAME": "SMITHSONIAN INSTITUTION"}, "INGESTDATE": "2014-07-31T11:14:53.630859, " "ID": "AF994D5BDBE589A9367E79980BFD7470, " "INGESTIONSEQUENCE": 14, "ISSHOWNAT"":HTTP://COLLECTIONS.SI.EDU/SEARCH/RESULTS.HTM?Q=RECORD_ID%3ASIRIS_SIL_986353&REPO=DPLA, " "SOURCERESOURCE": {"DESCRIPTION": ["TRADE LITERATURE, " "NEW YORK AGENCY, WILSON BROS. TOY CO., 119 CHAMBERS STREET., " "LITH. OF MILTON BRADLEY CO., SPRINGFIELD, MASS.\ "—P. [4] OF COVER, " "GAMES AND PUZZLES—SECTIONAL PICTURES AND MAPS—TOYS AND BLOCKS—NOVELTIES"], "LANGUAGE": [{"NAME": "ENGLISH, " "ISO639_3": "ENG"}], "FORMAT": ["55 P.: ILL., " "24 CM"], "@ID": "HTTP://DP.LA/API/ITEMS/AF994D5BDBE589A9367E79980BFD7470#SOURCERESOURCE, " "TEMPORAL": [{"BEGIN": "1889, " "END": "1889, " "DISPLAYDATE": "1889"}, {"BEGIN": "1889, " "END": "1889, " "DISPLAYDATE": "1889—90 I.E. 1889"}], "TITLE": "CATALOGUE OF GAMES, SECTIONAL PICTURES, TOYS, PUZZLES, BLOCKS AND NOVELTIES / MADE BY MILTON BRADLEY COMPANY, " "COLLECTION": [{"@ID"":HTTP://DP.LA/API/COLLECTIONS/9463ABD671D67157E760344619BFBB9C, " "ID": "9463ABD671D67157E760344619BFBB9C, " …

***Source*: Digital Public Library of America (excerpt of record)**[1]

A COMPARATIVE SETTING

In November of 2012, at the top of the Chattanooga Public Library—a concrete monument to an older era of collecting—I first caught a comparative view of data. I was in Chattanooga, Tennessee, for an "Appfest," the first of its kind hosted in support of the Digital Public Library of America (DPLA), introduced at the beginning of this book.[2] The DPLA was initiated only two years earlier, on October 1, 2010, by a coalition of computer scientists, lawyers, librarians, and philanthropists from across the United States who

3.1

The brutalist era building where the first DPLA Appfest was held. Courtesy of the Chattanooga Public Library.

convened with the goal of gathering widespread collections data into an integrated index of cultural history "dedicated to the public good."[3] Since then, the DPLA has enlisted a broad array of institutional collections and collaborators: private and public universities, national institutions like the Smithsonian, and elite collections like the Getty. It has also developed tools for public access and succeeded in attracting a broad community.

I spent two days in Chattanooga, working with early participants in this community. We sorted through data compiled by the DPLA on books, newspapers, maps, photographs, and other objects of cultural import in a hanger-scaled space that was as grand as the nascent aspirations of the DPLA yet also windowless, with little connection to Chattanooga and what it might have meant to converge in this place. Our work was intended to help liberate data from their origins so as to make them accessible anywhere. Instead, the experience nudged me to question the goal of separating data from their originating institutions of collecting. My encounter with the DPLA—an experience that stretched over the next few years—helped me to articulate the third principle that structures this book: *data are assembled from heterogeneous sources, each with their own local conditions*.

The previous chapter focused on one data setting, the Arnold Arboretum. And even there, differences in data formats and uses were evident. But what happens when data from different institutions are brought together? How might we jointly hold or reconcile their incongruous place attachments? When data from multiple settings are juxtaposed, as in the DPLA, there is an inevitable clash between discordant originating data cultures.

The DPLA is an example of a data infrastructure: a meta collection, which agglomerates digital resources from distributed sites of production.[4] Data infrastructures are often understood in terms that are seemingly technical. *Ingestion* grapples with how to collect and store heterogeneous data sets.[5] *Interoperability* considers how to make those data speak to one another.[6] *Enrichment* contends with how to add more information to ingested records.[7] And finally, *interface* has to do with how to make those records accessible and actionable.[8] But data infrastructures also awaken an ancient cultural ambition, as old as stories of Babel, to sever knowledge from its origins.

Using the DPLA as an illustrative case and harnessing a variety of techniques for local reading, I seek to uncover the culturally rooted place attachments that persist in data infrastructures. Reading DPLA data up close, with a focus on their classifications, schemata, constraints, errors, absences, and rituals, can reveal telling traces of their origins. Meanwhile, visualizing the encompassing data structures nested in the DPLA can offer us a glimpse of what is lost in practices of normalization: a process by which data are made to conform to an expected range of categories and values. For the DPLA, that format is its "MAP," an internally defined metadata application profile.[9]

Despite the DPLA's relatively low profile, the example demonstrates better than any other I know why data differ from one another. The data of the DPLA were created

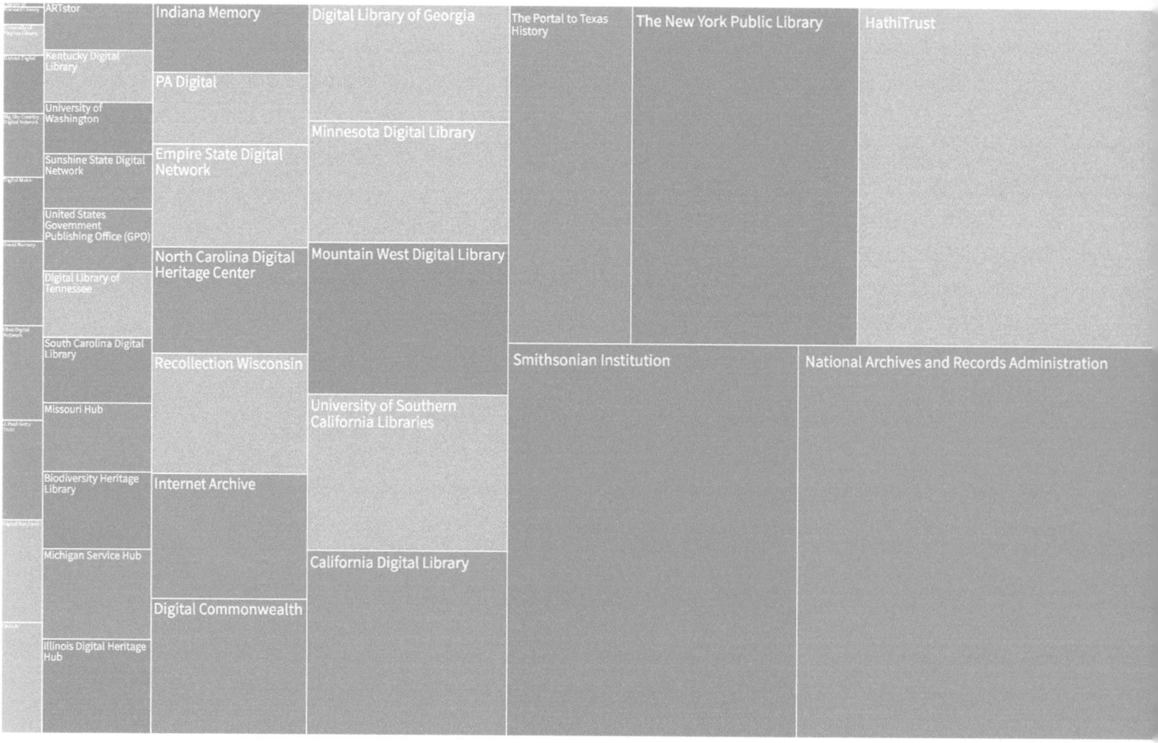

ARTstor
Kentucky Digital Library
University of Washington
Sunshine State Digital Network
United States Government Publishing Office (GPO)
Digital Library of Tennessee
South Carolina Digital Library
Missouri Hub
Biodiversity Heritage Library
Michigan Service Hub
Illinois Digital Heritage Hub
Indiana Memory
PA Digital
Empire State Digital Network
North Carolina Digital Heritage Center
Recollection Wisconsin
Internet Archive
Digital Commonwealth
Digital Library of Georgia
Minnesota Digital Library
Mountain West Digital Library
University of Southern California Libraries
California Digital Library
The Portal to Texas History
The New York Public Library
HathiTrust
Smithsonian Institution
National Archives and Records Administration

3.2

The Library Observatory evolved to incorporate new contributing collections to the DPLA. Image by the author, Matthew Battles, and Jessica Yurkofsky. See http://www.libraryobservatory.org.

as part of entirely different knowledge systems, some of which have evolved over decades at long-lived institutions like universities, museums, and public archives. Others were invented in seclusion in order to serve the owners of small private collections. Acknowledging these conditions means learning to engage data infrastructures not as large, homogeneous sources of information but rather as sites of controversy where varied conceptions of data come into conflict.[10] If we are to develop critical perspectives on data, I believe that the DPLA can help us learn to see both the forest and the trees.

TAKING DATA INFRASTRUCTURES APART

The Appfest, as its name suggests, was a minimarathon for designing software applications that would extend the DPLA, including rapid project pitches, informal tutorials, and plenty of head-down coding time. Such gatherings, more often called "hackathons," are meant to both kick-start new computing projects and foment community growth around a set of technical problems. They predate contemporary data infrastructures. In fact, the term *hack* comes from an earlier culture of computing, which Sherry Turkle defines in terms of its creative and sometimes subversive use of bricolage—an approach to making software that she compares to a conversation.[11] Today, hackathons still espouse this bottom-up, conversational approach to computing, but they are often focused on data and how to wrangle them for productive means. They are meant to redress, without overtly acknowledging, a pervasive problem: data infrastructures are frequently brought together with little concrete sense of how they should be used.

In Chattanooga, the Appfest began, as many do, with an introduction to the data. At the time, the DPLA had a much smaller trove, drawn from only three contributing institutions: the Digital Library of Georgia, Minnesota Digital Library, and South Carolina Digital Library. We were encouraged to produce projects that might enrich those initial offerings for a broad public audience. In considering the task from my burgeoning local perspective, I was struck by the fact that it was so hard to see the collection in terms of its parts. Could a panoramic view, of the sort introduced in chapter 2, enrich our understanding of the DPLA's current holdings? I was curious to learn who had contributed what and when. Since this view wasn't possible at the time, I went about trying to construct it, together with several other Appfest participants.[12]

Our project, later dubbed the "Library Observatory" by one of my collaborators, the writer and polymath Matthew Battles, is structured as a tree map: a hierarchical form of visualization composed of nested boxes for data that conceptually mimics the branching structure of a tree. Each box contains or represents all the entries at one level of the collection.

The largest boxes in the visualization, representing the contributing institutions of the DPLA, form the metaphoric base of the tree. In our initial design, the box containing entries from the Georgia library takes up half the overall image because it contributed that portion of the total DPLA holdings at the time. Minnesota and South Carolina are about equal. Inside each of these contributor boxes are smaller ones, labeled according to the next level of organization for each collection.

This is where their structures start to diverge noticeably. Georgia's subcompartments include "GA Obituaries," "1730-1842," "Civil Rights Digital Library," "Georgia Government Publications," and a large subcollection confusingly labeled the same way as the higher-level container, "Digital Library of Georgia." The South Carolina and Minnesota collections have their own subcollections, many of which also seem bemusing or simply unhelpful from an outsider's perspective; they are often named for local cities, institutions, or people that have little significance for a national or international audience.

Our crude map of these conditions was a mess, reflecting the discordant structure of the data contained therein. Nevertheless, it was a useful first step toward the goal of taking apart the DPLA in order to reveal its invisible localities. Such a map can provide a useful illustration of how a data infrastructure is composed. But how might a tree map be made into a useful navigation tool for visitors? When a search query is formed by a visitor, would it help them to know where, among the contributing collections, the results are returning from?

The Library Observatory project, which called attention to more problems than it solved, did not win any prizes at the Appfest.[13] Although antithetical to my agenda to reveal the heterogeneity of the DPLA, the winner that weekend, aptly named "Dedupe," is also helpful for understanding data infrastructures and the challenges they face. Dedupe treats the entire DPLA as if it were a one-dimensional data array, plagued by redundancies that need to be nulled.

Ironically, nowhere are the differences within a data infrastructure more apparent than in efforts to normalize them. The term *dedupe* is shorthand for an automated process that will rid the DPLA of duplicate entries, deemed repetitive by an algorithm. Yet the process of identifying seemingly identical digital versions of books, newspapers, and other collections objects can also reveal key differences that have the potential to illuminate what each object means in its originating context. Duplicates are a key to learning about the heterogeneity of data infrastructures.

Instead of ridding the DPLA of redundancies, why not learn from them? When seen in this way, the initiative presents an opportunity to use data infrastructures to study the production of data—raising important questions about the local histories of collecting across the country: Who makes data and why? What does *data* mean in different cultures of collecting? What kinds of values and assumptions can data hold?

Independently of the DPLA, a number of recent projects have shed light on the importance of duplicates in library collections. For example, a project called "Book Traces" at the University of Virginia implicitly demonstrates the relevance of duplicates by systematically documenting the unique attributes of individual, physical books. The authors of "Book Traces" invite us to explore how "old library books bear fascinating traces of the past."[14]

186 BALLADS AND OTHER POEMS.

"Thus, seamed with many scars,
Bursting these prison bars,
Up to its native stars
My soul ascended!
There from the flowing bowl
Deep drinks the warrior's soul,
Skoal! to the Northland! *Skoal!*"*
—Thus the tale ended.

* In Scandanavia this is the customary salutation when drinking a health. I have slightly changed the orthography of the word, in order to preserve the correct pronunciation.

THE WRECK OF THE HESPERUS.

It was the schooner Hesperus,
That sailed the wintry sea;
And the skipper had taken his little daughter
To bear him company.

Blue were her eyes as the fairy-flax,
Her cheeks like the dawn of day,
And her bosom white as the hawthorn buds,
That ope in the month of May.

The skipper he stood beside the helm,
With his pipe in his mouth,
And watched how the veering flaw did blow
The smoke now West, now South.

3.3

An 1891 book of poems and ballads by Henry Wadsworth Longfellow, annotated by Jane Chapman Slaughter. Image from the "Book Traces" project.

Over the course of the "Book Traces" project, which is ongoing, the authors have enhanced existing collections data to describe what they call "interventions" in thousands of books. "Readers wrote in their books, and left pictures, letters, flowers, locks of hair, and other things between their pages."[15] In one example (figure 3.3), a book of poems by Henry Wadsworth Longfellow is annotated with bittersweet memories of reading it with a lost friend.[16] The creators of "Book Traces" have also developed generalized methods for physically surveying a large number of volumes for such interventions. Following on their project, others have been inspired to explore interventions as evidence of the complex lives of physical books and thus grist for the mill of academic research. Devoney Looser, an English professor at Arizona State University, writes that "the items found in the books ... provide important information about circulation and authorship, and are of interest to critics, historians and biographers."[17]

But beyond identifying unique conditions in redundant copies of books for the purposes of narrowly defined scholarship, "Book Traces" demonstrates the educational merit of examining physical collections. Rather than trying to streamline the digitization and ingestion process, as the DPLA has sought to do, what can be learned by seeing that process as an opportunity for critical reflection and hence a point of entry for student engagement? The challenges of digitization can, when seen in the right way, become lessons on the working of knowledge systems like library collections. What the "Book Traces" project misses, and I am seeking to unravel, are the ways in which data can also act as cultural markers of past collection practices, and how they differ from one era or institution to the next. The agglomerated data of the DPLA provides a number of opportunities to understand what it means to look within data infrastructures for the local conditions in all data.

IDENTIFYING THE LOCAL

The DPLA became a self-supporting nonprofit a year after the Appfest. On its first anniversary in 2014, the organization reportedly contained "over 7 million digitized cultural heritage items from 1,200 contributing institutions across the United States." Today, in 2019, the DPLA is a thriving nonprofit organization, guided by a board of directors that includes academics, librarians, publishers, and businesspeople. The primary interface to the DPLA—a standard search bar with the heading "A Wealth of Knowledge: explore 11,578,169 items from libraries, archives, and museums"—promises equal access to each individual repository (figure 3.4). The initiative describes its mission of cultural collecting as all encompassing: "It strives to contain the full breadth of human expression, from the written word, to works of art and culture, to records of America's heritage, to the efforts and data of science."[18]

But a basic search of the DPLA's unified collections (figure 3.5) conceals the striking heterogeneity and unevenness of the underlying data. Below, I demonstrate local readings of collections data using examples drawn from the DPLA. These efforts variously reveal local classifications, schemata, constraints, errors, absences, and rituals, all of which are rooted in the contingencies of the places in which DPLA data are made.

Questioning Classifications

At the conceptual level, data are shaped by local classifications. Yet audiences typically don't take notice of them, unless their origins are unfamiliar. Jeffrey Licht, a technologist working for the DPLA, calls attention to a record that would appear unfamiliar outside South Carolina.[19] The DPLA contains a group portrait, contributed by Clemson University, with a field labeled "coverage" containing a single string, "upstate," presumably referring to a place in South Carolina. The field (coverage) and string (upstate), however, have little meaning to either Licht or me. Neither of us are from the region. But such language shouldn't be presented as anomalous: a mere obstacle to accurate geocoding, the process of turning places into coordinates on a map. Instead, the example should compel us to think about the local nature of all place-names. Local classifications are a product of geographies as well as other social boundaries, such as those that separate disciplines.

Seeing Schemata

In the collections data of the New York Public Library, a major civic institution and early contributor to the DPLA, one can find at least 1,719 unique date schemata: ways of recording the moment that a book, image, or other library artifact came into the world. This detail was already introduced at the beginning of the book, but below I offer a deeper look at a sample of abstracted date schemata. These are not actual dates. Rather, each represents one way of documenting a date of publication.

Printed by Thomas; Badger, Jun (1)
pref ____] (1)
__ March, ____ (3)
probably before ____ (7)
[c____]/____ (130)
____-____, ____-____ (209)
____-____, reissued through ____ (240)
____-__-__/____-__-__ (438)
ca. ____'s (640)

The "_" in each schema is a variable standing in for a variety of possible integers. The number in parenthesis indicates the total times that the format appears in the New York Public Library catalog. Thus the common schema ca. ____'s, used 640 times, might be encountered as ca. 1950s. The less common formats are at the extreme ends of uncertainty: either highly ambiguous or strangely specific. In one case cited here, the format includes the name of the printer. It appears only once. Although we can't

BROWSE BY TOPIC BROWSE BY PARTNER EXHIBITIONS PRIMARY SOURCE SETS | ABOUT DPLA NEWS | DPLA PRO
D P L A
DIGITAL PUBLIC LIBRARY OF AMERICA
Donate
Discover 21,455,806 images, texts, videos, and sounds from across the United States
Search the collection
Search
Browse by Topic
New? Start Here

Online Exhibitions
Browse all Exhibitions >

CHISHOLM PRESIDENT

Two Hundred Years on the Erie Canal
American Empire
Battle on the Ballot: Political Outsiders in US Presidential Elections
Race to the Moon

Primary Source Sets
Browse all Sets >

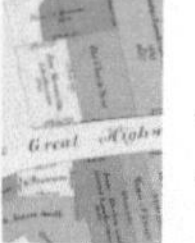

Cotton Gin and the Expansion of Slavery
Elie Wiesel's Night and the Holocaust
Victorian Era
Immigration through Angel Island
Dutch New Ne

3.4

The main page of the DPLA website.

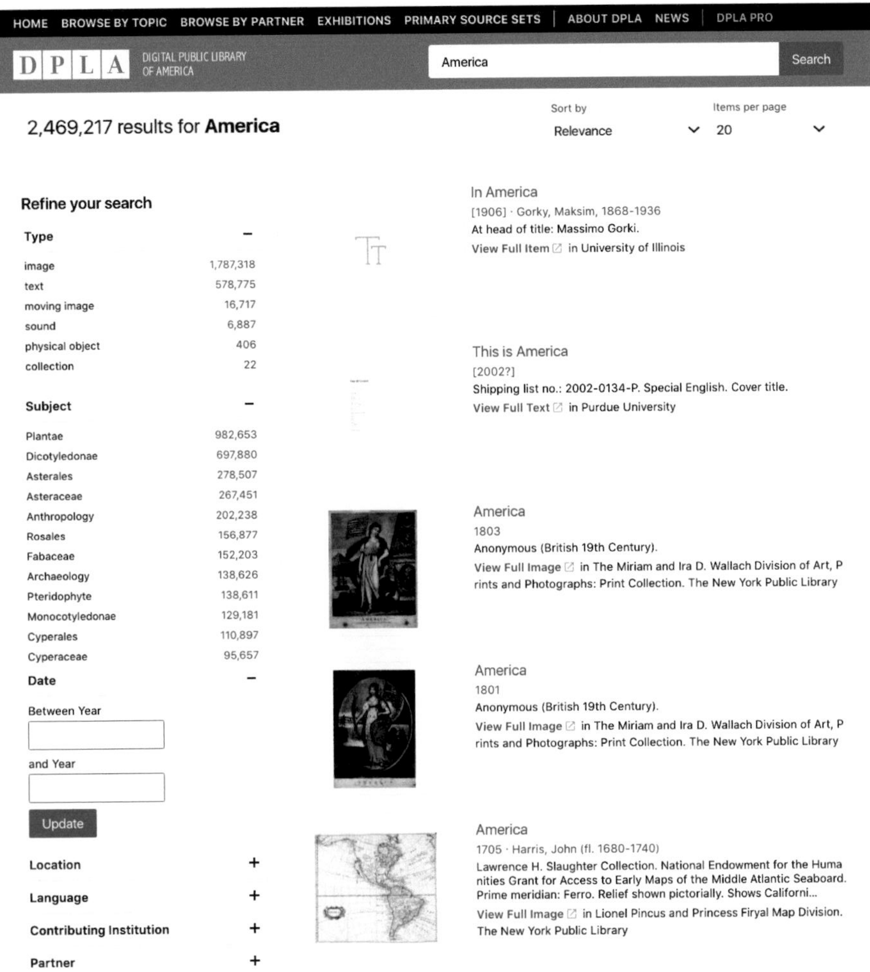
HOME BROWSE BY TOPIC BROWSE BY PARTNER EXHIBITIONS PRIMARY SOURCE SETS | ABOUT DPLA NEWS | DPLA PRO
D P L A DIGITAL PUBLIC LIBRARY OF AMERICA
America
Search
2,469,217 results for America
Sort by
Relevance
Items per page
20
Refine your search
Type
image 1,787,318
text 578,775
moving image 16,717
sound 6,887
physical object 406
collection 22
Subject
Plantae 982,653
Dicotyledonae 697,880
Asterales 278,507
Asteraceae 267,451
Anthropology 202,238
Rosales 156,877
Fabaceae 152,203
Archaeology 138,626
Pteridophyte 138,611
Monocotyledonae 129,181
Cyperales 110,897
Cyperaceae 95,657
Date
Between Year
and Year
Update
Location
Language
Contributing Institution
Partner
In America
[1906] · Gorky, Maksim, 1868-1936
At head of title: Massimo Gorki.
View Full Item in University of Illinois
This is America
[2002?]
Shipping list no.: 2002-0134-P. Special English. Cover title.
View Full Text in Purdue University
America
1803
Anonymous (British 19th Century).
View Full Image in The Miriam and Ira D. Wallach Division of Art, Prints and Photographs: Print Collection. The New York Public Library
America
1801
Anonymous (British 19th Century).
View Full Image in The Miriam and Ira D. Wallach Division of Art, Prints and Photographs: Print Collection. The New York Public Library
America
1705 · Harris, John (fl. 1680-1740)
Lawrence H. Slaughter Collection. National Endowment for the Humanities Grant for Access to Early Maps of the Middle Atlantic Seaboard. Prime meridian: Ferro. Relief shown pictorially. Shows Californi...
View Full Image in Lionel Pincus and Princess Firyal Map Division. The New York Public Library

3.5

An example search on the DPLA website.

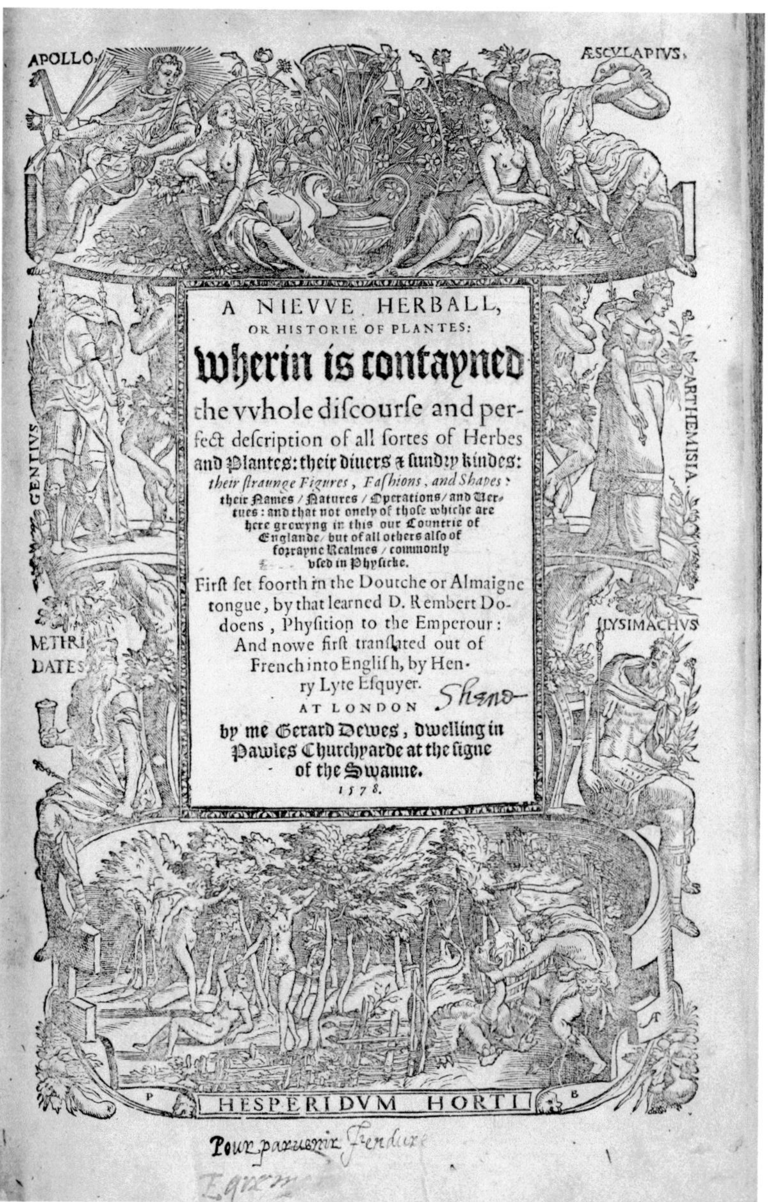

APOLLO. ÆSCVLAPIVS.

A NIEVVE HERBALL,
OR HISTORIE OF PLANTES:

wherin is contayned
the vvhole discourse and perfect description of all sortes of Herbes and Plantes: their diuers & sundry kindes: their straunge Figures, Fashions, and Shapes: their Names / Natures / Operations / and Vertues: and that not onely of those whiche are here growyng in this our Countrie of Englande, but of all others also of forrayne Realmes / commonly vsed in Physicke.

First set foorth in the Doutche or Almaigne tongue, by that learned D. Rembert Dodoens, Physition to the Emperour: And nowe first translated out of French into English, by Henry Lyte Esquyer.

AT LONDON
by me Gerard Dewes, dwelling in Pawles Churchyarde at the signe of the Swanne.
1578.

GENTIVS. ARTHEMISIA. METHRIDATES. LYSIMACHVS.

HESPERIDVM HORTI

3.6

An example of a historical book that causes numerous production artifacts from the DPLA collections.

understand the most obscure date schemata without further inquiry, they stimulate curiosity about their own invisible local histories.

Attending to Constraints

Ways of inscribing data are always constrained by local conditions. One well-known example of the technical limits on data comes from the turn of the millennium. Leading up to the year 2000, digitized date codes had to switch from two to four digits. Otherwise "00,01,02" might be mistaken for 1900,1901,1902 versus 2000, 2001, 2002. Across contributing collections to the DPLA—and indeed the world—databases had to be updated at great cost. And still, fears persisted that some unseen conflict in formats might cause whole systems to fail. Two-digit date formats are local markers of a bygone era. In previous eras, when storage space was much more costly, programmers used two-digit date codes to save space. We should read such legacy constraints as evidence of the way that data are located in a technological moment. But the lesson is more general: without the software and hardware of their era as well as operating knowledge thereof, data would not be accessible at all.

Decoding Errors

Every contributing collection to the DPLA contains errors. One doesn't need much instruction to notice misspelled words or misplaced punctuation. But it takes a degree of local knowledge to see that such errors are not random. In fact, we can unpack them as evidence of localized cataloging practices.[20] They stem from situated processes of data production.

Badly scanned text, blurred photos, and moiré effects—all common in DPLA records—are a result of specific imaging technologies and ways of making use of them within a local setting. Typographic errors sometimes originate in the use of type from a particular historical moment (figure 3.6). They are brought on by the optical character misrecognition of unusual typefaces, ligatures, or unexpected characters. For instance, the standing *s* in early modern English typography is routinely mistaken for an *f* by character recognition systems. Understanding this has value beyond the amusement of contemporary readers. Alternatively, errors can be brought about when content is mistaken for code. Brackets, dollar signs, and semicolons can be interpreted as instructions to be carried out by a computer program.

We often hear about the arduous but imperative need to rid data sets of such flagrant errors through acts of cleaning or filtering. As I first explained in chapter 2, however, such instances of data dirt are simply out of place. In other words, errors in collections data might be better understood as signifiers taken out of their original interpretative contexts. We should learn to read data dirt as important traces of their own local production.

Revealing Absences

More subtly than in the previous illustrations, data are defined by what they leave out. Former Smithsonian historian Marya McQuirter recounts having searched her institution's catalog, one of the largest contributors to the DPLA, for the terms *black* and *white* (figure 3.7).[21] The first brings up lots of examples of African American artists; the museum diligently documents work created by or about people who identify racially as black. But the search term *white* brings up little about race, other than the occasional piece linked to white supremacy. White is not a racial identity that the Smithsonian typically tracks. Instead, the category exists as an absence that reveals a bias.[22] Whiteness is not critically examined by the institution. Yet it is the racial identity of the vast majority of artists whose work is shown at the Smithsonian. As this example demonstrates, absences are part of deeply rooted systems of representation—in this case, white supremacy—reified in data.

Observing Rituals

Finally, and less overtly visible, are the local rituals that shape data. I use the term *ritual* here to identify cultural practices with data that have their own significance as symbolic expressions or community-making activities.[23] Thomas Ma, a cataloger at the Harvard Library—another of the heaviest contributors to the DPLA—reflects on the way that cataloging has changed over the course of his career.

> I remember when I first started at the law school, I was told by the person in charge of technical services that "you weren't worth anything if you didn't have a backlog." And nowadays it's like if you have a backlog, you have the cooties. So the backlog [used to be] evidence of a certain kind of care and quality and attention in the catalog processing. And now the backlog is a distinct liability.[24]

In this instance, practices with data are closely tied to professional identity and status. Moreover, the change in cataloging practices has significant practical implications. For when backlogs pile up—sometimes consisting of tens of thousands of accumulated books that need to be processed—cataloging is outsourced. "We used a company in Arkansas. I guess they find cheap labor. ... They just grab people off the street and say, here, slap a record together and move it on."[25] Ma's words are laden with an implicit argument for preserving the social milieu in which he was trained. His data rituals are evidence of a local social order. Indeed, without the proper rituals, argues Ma, the quality of library data is in danger. Ma forecasts a dark future for collections: they contain more entries than ever before, but their data—the maps to those collections—are increasingly thin. From this perspective, the movement toward data infrastructures can paradoxically make individual records less accessible.

As the examples above illustrate, there is no such thing as universal data. Data are situated within the means of their production, the infrastructures required to maintain them, their systems of representation, and the social order that they reproduce. This perspective extends social studies of information that focus on the specificity of data at the level of the institution, such as the museum or laboratory.[26] Such studies often assume that standards are the primary forces that shape data. Lisa Gitelman writes that "every discipline and disciplinary institution has its own norms and standards for the imagination of data."[27] As the above readings of DPLA data suggest, though, local variations can also be subject to a number of historical, technological, and cultural contingencies, which transcend disciplinary boundaries. Too often, these differences are passed off as anomalies, to be resolved by normalizing data. But we can learn to see differences in data as markers of otherwise-invisible local conditions that must be understood for meaningful analysis.

I do not mean for the six features examined above—classifications, schemata, constraints, errors, absences, and rituals—to be taken as a fixed typology for local readings of data. Rather, they convey the contingent character of data through examples that cover a range of possible scales and local ties. What appears to be local in data depends on emergent differences among data aggregated from many places. As first suggested in the introduction yet more fully illuminated here, the local is only intelligible when seen through a comparative lens. When data are drawn together from disparate origins, conflicting practices of data production are suddenly apparent.

Local markers are especially relevant when exploring big data.[28] For although the term indicates a departure from the local, the rise of the big data phenomenon has ironically made the local qualities of data more significant. Under big data, distributed records with discordant local ties are increasingly estranged from their creators and presented to audiences other than those first intended.

The DPLA might not conform to the strict definitions of big data, mentioned earlier in the book, emerging from critical data studies: high in volume, variety, and velocity. Yet it and other data infrastructures explored in this book are important references for grappling with the encompassing cultural phenomenon of big data, which is best understood as a desire for data sets that are intended to be comprehensive and autonomous, capable of yielding insights without contextual information.[29] This definition speaks to the universalizing ambitions of many data infrastructure projects, and prompts us to think about when and why they fall short.

VISUALIZING THE LOCAL

Reading the data of the DPLA one entry at a time, and looking for local instances of schemata, errors, constraints, classifications, absences, and rituals, can be revelatory but also time consuming. Furthermore, some of those conditions are instances of patterns

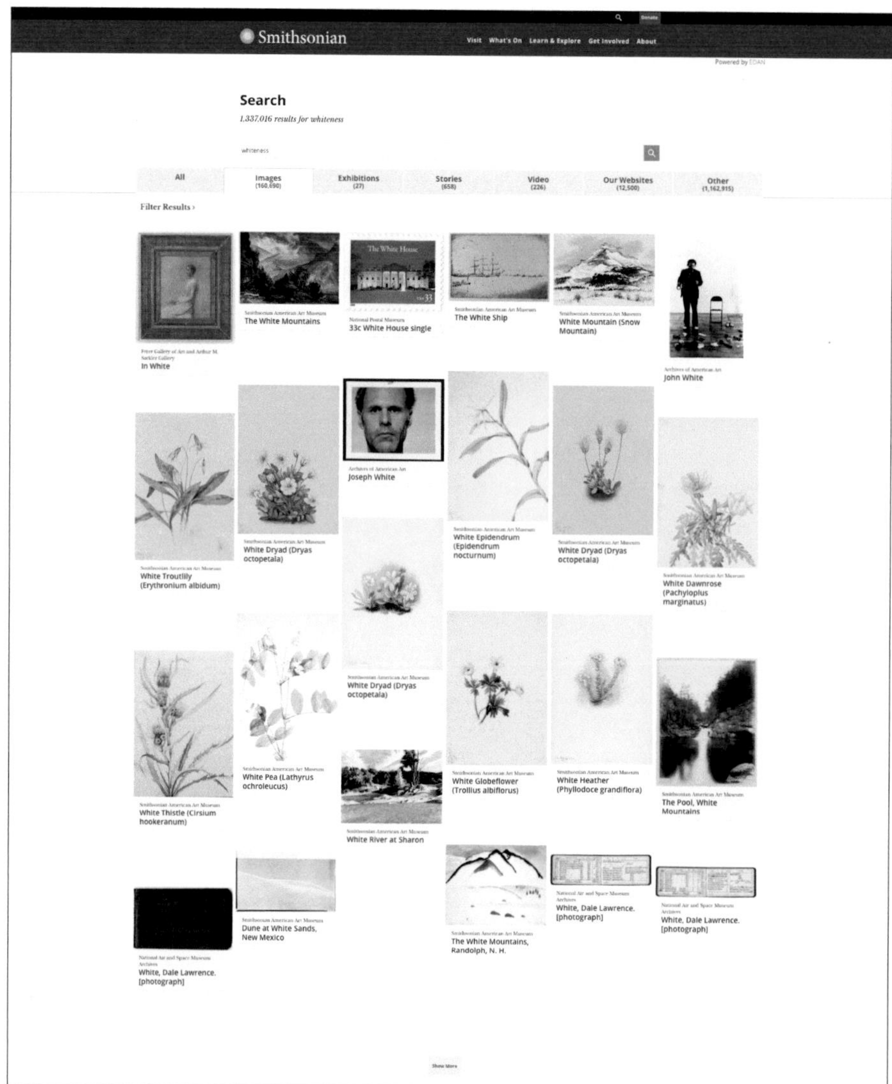

3.7

Example searches conducted by the author for *whiteness* and *blackness* on the Smithsonian website.

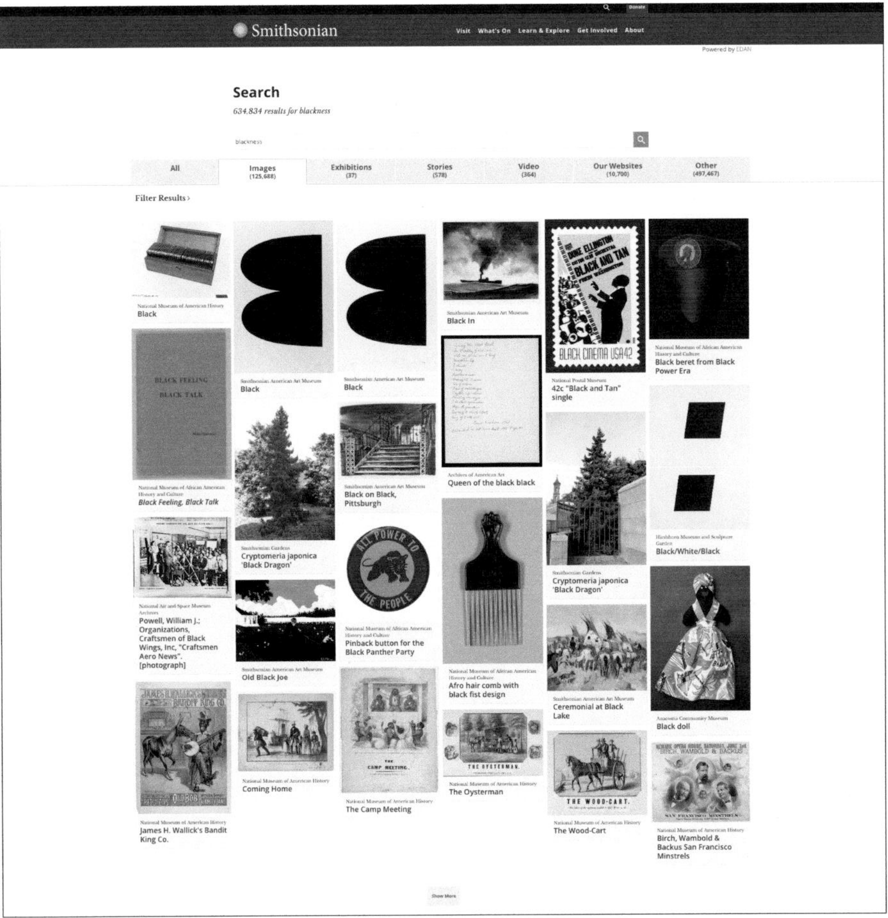
Smithsonian
Visit
What's On
Learn & Explore
Get Involved
About
Donate
Powered by EDAN
Search
634,834 results for blackness
blackness
All
Images (125,688)
Exhibitions (37)
Stories (578)
Video (364)
Our Websites (10,700)
Other (497,467)
Filter Results ›
Black
Black
Black
Black In
42c "Black and Tan" single
Black beret from Black Power Era
BLACK FEELING BLACK TALK
Black Feeling, Black Talk
Queen of the black black
Black on Black, Pittsburgh
Cryptomeria japonica 'Black Dragon'
Cryptomeria japonica 'Black Dragon'
Black/White/Black
Powell, William J.; Organizations, Craftsmen of Black Wings, Inc, "Craftsmen Aero News". [photograph]
ALL POWER TO THE PEOPLE
Pinback button for the Black Panther Party
Old Black Joe
Afro hair comb with black fist design
Ceremonial at Black Lake
Black doll
James H. Wallick's Bandit King Co.
Coming Home
THE CAMP MEETING.
The Camp Meeting
THE OYSTERMAN.
The Oysterman
THE WOOD-CART.
The Wood-Cart
BIRCH, WAMBOLD & BACKUS
SAN FRANCISCO MINSTRELS
Birch, Wambold & Backus San Francisco Minstrels
Show More

that appear at larger scales. Revealing the larger pattern and its causes requires different methods of data presentation. Data visualization, first utilized in the last chapter as a means of seeing the forest, all the data at once, can also illuminate the trees, markers of difference within. Used in this way, visualization can become a mechanism for "infrastructural inversion," through which foundational though invisible patterns are made visible.[30]

When Geoffrey Bowker and Susan Leigh Star introduced the concept of infrastructural inversion in *Sorting Things Out*, they proposed that the most straightforward way to make infrastructure visible is to break it. Indeed, errors in DPLA data, such as the misrecognition of the long *s* by optical character recognition, can call our attention to the infrastructural processes by which those data are produced. Yet there are other strategies for seeing infrastructures that we might invoke. Here, as in previous examples, I use a comparative approach. Comparative visualization can illuminate how locally distinct collections are assembled and assimilated differently in data infrastructures like the DPLA. I developed two distinct visualization programs to demonstrate this at different scales.

The first of the two, entitled "A Comparative Visualization of Temporalities in DPLA Data," or "Temporalities" for short, revisits the startlingly distinctive date formats to be found in the ingested collections data of the DPLA. These are data that have already been normalized: changed in order to conform to a single format. But wisely, the DPLA chose to preserve the original records. This visualization unearths those records and organizes them from most legible to least, but by not human legibility. Their machine legibility is determined by the prevalence of date signifiers like months, days, and years, which a computer can easily be made to identify.

This way of putting the date formats on the page gives the reader a sense of the range of original schemata drawn together in the DPLA as well as the normalization work necessary to make them conform to a single format. The temporalities that do not conform are subject to procrustean measures: stretched out or cut short as needed. This is the evidence of what must be obscured or added in order to normalize just one data field.

A Comparative Visualization of Temporalities in DPLA Data

Temporalities (figure 3.8) is coded in JavaScript to circumvent the DPLA's default web interface (www.dp.la/) and communicate directly with its application programing interface (API).[31] This backdoor interface is the only means of accessing the original records from contributing institutions like the New York Public Library, Smithsonian, or Digital Library of Georgia.

Here is a brief explanation of how the program works. It begins with a search request to the API. I have used the search term *America* in the trial run above as a gesture to the full scope of resources that the DPLA aims to encompass. The results of this search, returned in JavaScript object notation, are at first stored in memory. For each individual record returned by the API, the code identifies a field titled "sourceResource. date.displayDate." This is the yet-to-be-normalized date format provided to the DPLA by contributing institutions. Other fields found in each DPLA record, date.begin and date. end, contain normalized date values: representations of the date that have been created to facilitate consistent searches from the main DPLA interface.

The code then reformats each sourceResource.date.displayDate: integers (frequently used to represent days, months, and years) are converted to underscores (\\d to _), and months are converted to double underscores (Jan|Feb|Mar| ...) to __). Integers and months are the most common features in these dates. Seeing their arrangements provides a useful comparison of underlying commonalities. Finally, each format is saved to a list, used to keep a tally of how many instances of each common string (or schemata) are found. The final visualization shows this list: all the date formats saved for that one search.

In order to help the reader make sense of this list, it is sorted in terms of its machine legibility—defined here by the ratio of underscores to other characters. This ratio indicates the amount of machine-readable information in the format. When examples have the same ratio, those with more underscores are listed first. Date formats that are text heavy and thereby less machine readable will be pushed to the bottom. These dates are more difficult to normalize. The hardest dates to normalize, the strangest or most unexpected ones, require manual translation. For example, Roman numerals cannot easily be read by a machine. A program is likely to interpret all Vs, such as the V in "Version," as the Roman numeral equivalent of 5. Data cleaning has its limits, or at least requires a lot of locally sensitive rules. The numbers beside each format indicate how many times each format appeared in records returned by the API search.

Temporalities inverts the infrastructural process of ingestion. It does so by sectioning the DPLA along a single field and exposing traces of the deep intricacy of contributed collections. In Temporalities, that intricacy is exhibited in the range of localized date formats to which data are subject across cultures of collecting. In the context of the DPLA, meant to be an inclusive repository of US cultural history, such variations in data are worthy of attention and even beautiful in their own right. The second program, entitled "The Shape of DPLA Data," or "Data Shapes," also grapples with this heterogeneity and how to present it, but with a focus on the entire data structure of each DPLA entry.

3.8

The Temporalities application visualizes date formats from the DPLA in order of their machine legibility. Image by the author and Peter Polack.

Count	Format
31	________
1	_____
16110	____
18	________-________
2	____-________
827	____-____
404	____/____
1	____–____
1	____,____
2301	____-__
23	__ ____
12	____/__
11	____–__
3	____ __
2	____-_
12	____-__/____-__
11	____-__-__-____
2	____-__-____-__
1	____-____[____]
3597	____-__-__
16	__/__/____
13	____-____]
5	____-____?
5	____, ____
4	c____-____
4	____ __ __
3	__-__-____
3	____, c____
2	____/____?
2	____--____
2	____]-____
1	____-__ __
1	____-__/__
580	____]
347	c____
168	____-
29	____?
12	__ __
11	-____
8	____~

3 ____,

3 ©____

2 ____s

1 d____

64 _/__/____

9 ___u-____

4 __/_/____

2 ____-___u

2 ____-_-__

1 _ __ ____

1 ____-__-_

1 ____ __ __-__

12 ____-__-__/____-__-__

4 ____-__-__-____-__-__

1 ____-____, c____-____

3 ____, c____-____

8 ____-__]

6 _/_/____

2 ____]-__

2 __. ____

1 c____-__

1 ____-__

1 ____-__-

1 _/__/___

1 ____-__~

1 ____-'__

1 ____-__?

1 __ -____

8 ___u

2 l___

1 ___-

6 ____-__-__T__:__:__

466 ____ - ____

108 ____, c____

20 [____-____]

11 ____-[____]

8 [____,____]

7 ____?/____?

7 __ __, ____

6 ____[c____]

4 ____ [____]

2 c____, ____

2 ____, ©____

2 ____, __ __

2 [____]-____

1 c____-____]

1 ____~/____~

1 ____?-____]

1 ____-____?]

1 ____[-____]

1 ____?-____?

210 ____-__-__T__:__:__-__:__

3 __ __-__, ____

13 _/__/__

1 _-__-__

1 ____s (____-____)

6 ____-__-__T__:__:__Z

5 __ _, ____

1 _d __ ____

3 __ __-__ _, ____

1 {____-__-__..____-__-__}

2 ____-__ [c____-__]

3 ____ [c____-__]

1 ____-__ [c____]

110 [____..____]

57 ____ [c____]

17 c. ____-____

5 [____, ____]

3 ____ to ____

3 [c____-____]

2 ____, [____]

2 ____ [©____]

2 ____,[c____]

2 ____ or ____

2 [____,c____]

2 [____?]-____

1 ____, c.____

1 [____]-c____

1 c[____-____]

1 __ __y. ____

1	____-[c____]
1	[____?-____]
1	____, c____-
1	__th __ ____
1	c____-c____]
1	[____], ____
1	____-[____?]
1	[____-____?]
24	__uu-____
10	____-__uu
9	[____-__]
5	____-[__]
3	___u-___u
2	[____]-__
1	c____-__]
1	____, '__
1	[__]-____
1	____-__.]
1156	[____]
220	____?]
42	c____]
12	____.]
8	c____-
5	____?-
2	c.____
2	____--
1	-____]
1	c____"
1	_/_/__
1	-c____
1	____.,
1	[____-
1	c____?
1	____?>
1	_-_-__
1	____ (
1	____-____ [v._ ____]
1	____ [c____]-____ [c____]
1	_d __, ____
1	____ [____ or ____]

20	ca. ____-____
12	[____, c____]
6	____, [c____]
2	[____]-[____]
2	____ or ____]
2	____-ca. ____
1	____ [c.____]
1	c____-[c____]
1	[c____-c____]
1	[____?-____?]
1	[____?], ____
1	____ and ____
1	[____? c____]
1	[____ [c____]
1	[____?,c____]
1	[c____, ____]
16	[c____]-__
10	[c____-__]
1	c____-c__]
1	[____?-__]
3	___-]
1	___-?
3	[____ or ____]
2	____ i.e. ____
1	[____?, ____?]
1	[____, '__-__]
1	____, [c__-__]
1	____? to ____?
587	[____?]
452	[c____]
37	ca ____
9	c. ____
8	[©____]
4	[____]-
2	[____s]
2	____?-]
1	c[____]
1	____ AD
1	[____!]
1	____-]

1 [____.]

1 __[__]-

1 c ____-

1 wc____]

1 ____?]

1 -[____]

1 [c____-__] v._, ____[c____]

1 ____-__, t.p. ____

1 ____ [i.e.____]-__

3 ___u-__uu

2 _uuu-____

2 __uu-___u

3 ____-__ [v. _, ____]

3 [c____-c__]

1 [©____-c__]

1 [__-?]-____

8 circa ____-____

5 ____-circa ____

1 ____, t.p. ____

1 ____,repr. ____

1 ____, i.e. ____

1 ____- [v._, ____]

6 circa __ ____ - __ ____

1 ____ [i. e. ____-__]

11 ____ [i.e. ____]

1 ____ [cop. ____]

1 [etc.] ____-____

111 ca. ____

3 ca ____s

3 ____. --

3 ____, ca

2 [c.____]

2 [____-]

2 [____?]-

1 [c____.]

1 [____?-]

1 [c ____]

1 [c____]-

1 cop.____

1 [c____-]

1 c. ____s

1 [©____-]

1 [c____?]

1 [__ -__]

12 ___-?]

9 [___-]

2 [___?]

1 [l___]

9 __uu

2 __--

1 Vol. _-_ (____-____)-

1 ____-__, [v. l, ____]

3 ca. ____-ca. ____

1 modeled ____-____

1 ____, [i.e. ____]

1 ____ [cover ____]

1 Circa __ __, ____

1 Vol. __ (__. _, ____ to __. _, ____)-v. __ (__. ____ to __. ____)

1 ____-between ____ and ____

1 after __ ____

1 Circa __ ____

2 __. _st ____ [i.e. ____]

1 Vol. _ (__. _, ____ to __. _, ____)-v. __ (__. _, ____ to __. _, ____)

1 ____-____, [reprinted ____]

1 ____, i.e., [____]

5 __uu-__uu

2 ca. ____?

1 _uuu-___u

1 c____. --

1 c. ____s?

1 cop. ____

1 ca. ____s

1 ca. ____]

1 __r. _th ____ [i.e. ____]

1 after __, ____

26 [___-?]

1 [___-]-

1 __. ____-v. _, issue _ (__. ____)

4 ____, ____ printing

1 __th ed. (__ __/__)-__th ed. (__ __/__)

1 ____- [v. _, pt. _, ____]

1 Began with ____-____

1 ____; reprinted ____

1 __th ed. (__. __/__)

17 circa ____

10 [ca. ____]

7 after ____

6 [cop.____]

3 Circa ____

1 ____ circa

1 pref. ____

2 __--?

2 __--]

1 __nd ed. (__. __/__)-__th ed. (__. __/__)

1 _th ed., __. ____-

1 Vol. _ (__. ____)-vol. _, no. _ (__/__ ____)

1 __th ed. (__ _/__)-__th ed. (__. __/__)

1 Vol. _, no. _ (__/__ ____)

3 between ____ and ____

1 ____ [reprinted ____]

1 __th ed. (__. __/__)-

1 __th ed. (__. __/__)-__th ed. (__. _/__)

1 ca. ___-

1 between ____ and ____]

2 before ____

2 pref. ____]

1 [p.d. ____]

1 [cop. ____]

1 _d print. ____

2 ____, probably __

1 __th ed. (__. _/__)-

47 [between ____ and ____]

1 Egyptian________SIU.tif

1 Vol. _, no. _ (__. _, ____)-Vol. _, no. __ (__. _, ____)

1 Began in: __. ____

1 Began with __ ____

7 [pref. ____]

1 [cop. ____.]

2 _uuu-__uu

1 ___u-uuuu

3 __--?]

3 [__--]

2 [__-?]

1 Vol. _, issue _ (__ ____)-v. _, issue _ (__. ____)

1 Began with: __ ____

1 Print began and ceased with: No. _-____ (__. _, ____-__ __, ____)

1 patented ____

1 ____ printing

1 anno XI--____

1 [pref. ____.]

1 Egyptian________Rencher.tif

1 Egyptian________Antoine.tif

1 Vol. __, issue _ (__. ____)-v. __, issue _ (winter ____)

1 reprint ____, updated ____, original ____]

1 Began in ____?

1 ____ and later

1 Began in: ____

4 [__--?]

1 [__--]-

1 __--?]-

1 Began with _st ed., __. __, ____

1 Began with __th ed., __. _, ____

1 ceased with _th ed., __. __, ____

1 Colonial period, ca. ____-____

1 Began with _st ed., __ _, ____

5 Began with ____

1 MDCCLXXI [____]

1 [foreword ____]

2 Began with: ____

1 [__--?]-

2 _uuu

1 Circa early ____s

1 possibly ca. ____

1 [not before ____]

1 __th–__th century

1 MDCCLXXXIII [____]

2 Began in the ____s?

2 ____, printed later

1 probably after ____

1 Print began in ____

1 [etc., etc., ____?]

1 Yale University Press ____-__

1 Publication began with v. _, no. _ (__. __, ____)

1 Yale University Press ____-__

1 Print began in ____?

1 Vol. _-v. _

1 Ceased with v. __, issue _ (winter ____)

4 Printed in the year ____

2 __th century

1 Unpublished (before ____)

1 Print began with: _st ed. (____)

1 Began and ceased with: ____

2 late __th–early __th century

1 Print began with: Winter ____

1 Ceased, per publisher, in ____

2 late __th century

1 Print began with: _st ed., published in ____

1 Printed by Thomas Badger, __

1 -

1 n.d

7 uuuu

5 null

1 n.d.]

14 unknown

5 Unknown

3 undated

1 Undated

1 C. Tilt,

1 M.DCCC.l

1 [undated]

1 unpublished

1 C.T. Dillingham,

23 [Date Unavailable]

1 National capital press,

1 Cambridge Botanical Supply,

1 Printed by Gillespie Bros.,

1 The Womens social & Political union,

1 [Press of National printing co., etc.]

The Shape of DPLA Data

The second program, Data Shapes (figure 3.9), also begins with an API query to the DPLA. It returns a visualization for each individual data structure in the resulting DPLA entries. Each entry is represented as a network graph, which displays all the individual nested fields included therein. Related fields are connected by green threads and grouped at different levels in the data hierarchy within translucent convex shapes. Fields from the original contributor record are juxtaposed with fields created by the DPLA.

Like Temporalities, Data Shapes is an expressive project that examines what visualization can show us about the heterogeneity of the data infrastructures. Whereas most data visualizations focus on global patterns across a data set, Data Shapes show patterns of difference. It visualizes data fields and their relationships as opposed to data values, which is more typical of conventional visualizations.

Data Shapes interrogates the hierarchical structure of each DPLA entry as follows. The code starts at the root of the JavaScript object notation returned by the API. It then recurses through the entire structure of nested fields using a depth-first search. This means that for each (parent) field, the program visits all of its (children) subfields, then its (children's children) sub-subfields, before moving to the next (parent) field. For every individual (child) field on the tree, a new node is visualized and linked to the (parent) field one level up.

Data Shapes reveals the otherwise-invisible structures inside each record, including the original data from the contributing collection, and the Dublin Core data created for the DPLA.[32] The pink areas show different subcollections of data, and the green lines show how they are related. The bold numbers are indexes for fields used in the records and listed in a legend on the far left of the screen. The light numbers show the frequency of the field's use in the current image. This legend can help an informed reader decipher the original data from their DPLA-generated counterparts, formed during the process of ingestion. Taken together, Temporalities and Data Shapes suggest new ways in which visualization can help us see data rather than seeing through them. In collections of cultural history for our digital age, even data deserve to be engaged on cultural terms.

In addition to seeking out the cultural histories of data, we should acknowledge the histories behind data infrastructures. For such infrastructures are no more neutral than the data they contain. Since their earliest imaginings, information processing infrastructures were meant to draw data together, regardless of their origins, and mine them for insights.

3.9

The Data Shapes program visualizes the structure of individual JavaScript object notation records from the DPLA. Image by the author and Peter Polack.

0	24	0
1	1	context
2	1	dataProvider
3	1	admin
4	1	object_status
5	1	originalRecord
6	1	controlfield
7	37	#text
8	23	tag
9	12	1
10	5	2
11	4	3
12	4	4
13	4	5
14	2	provider
15	5	id
16	5	name
17	2	_id
18	1	datafield
19	17	subfield
20	31	code
21	17	ind1
22	17	ind2
23	2	6
24	2	7
25	1	8
26	1	9
27	1	10
28	1	11
29	1	12
30	1	13
31	1	14
32	1	15
33	1	16
34	1	leader
35	1	object
36	1	aggregatedCHO
37	1	ingestDate
38	2	type
39	1	ingestionSequence
40	1	isShownAt
41	1	sourceResource
42	1	publisher
43	1	description
44	1	language
45	1	iso639_3
46	1	title
47	1	format
48	1	rights
49	1	contributor
50	1	creator
51	1	extent
52	1	spatial
53	1	country
54	1	date
55	1	begin
56	1	end
57	1	displayDate
58	1	specType
59	1	identifier
60	1	subject
61	1	ingestType
62	1	score
63		hasType
64		collection
65		relation
66		stateLocatedIn
67		physicalDescription
68		tmp_high_res_link
69		tmp_image_id
70		usage
71		namePart
72		valueURI
73		authority
74		displayLabel
75		titleInfo
76		supplied
77		tmp_rights_statement
78		relatedItem
79		note
80		17
81		18
82		tmp_item_link
83		originInfo
84		dateIssued
85		place
86		placeTerm
87		location
88		shelfLocator
89		physicalLocation
90		version
91		genre
92		rightsStatementURI
93		schemaLocation
94		typeOfResource
95		topic
96		geographic
97		stringValue
98		keyDate
99		encoding

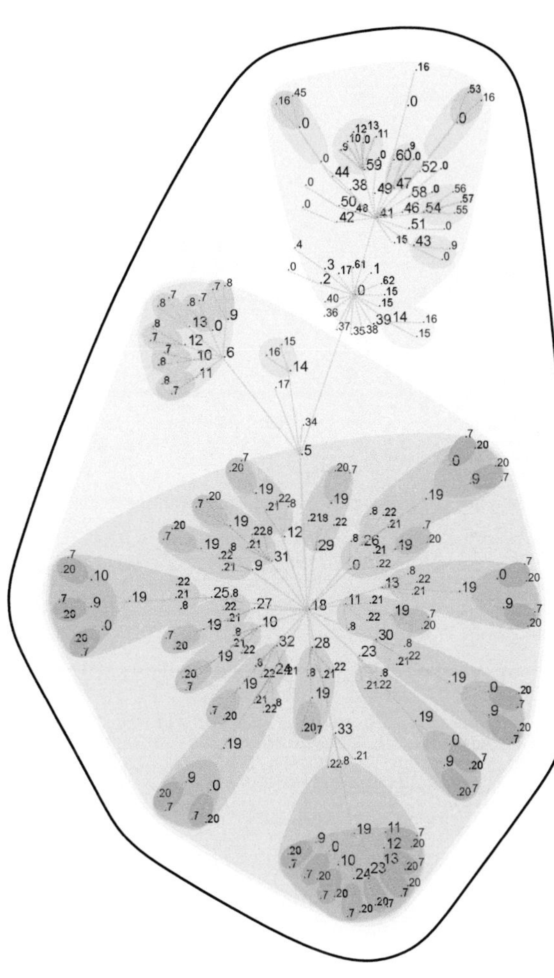

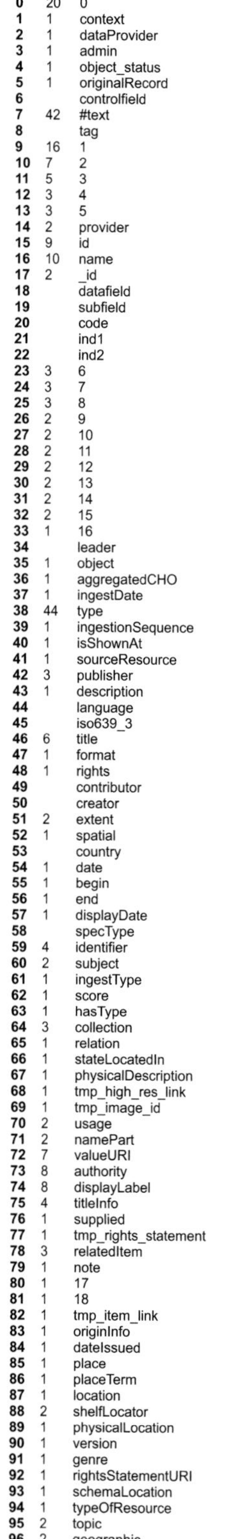

0	20	0
1	1	context
2	1	dataProvider
3	1	admin
4	1	object_status
5	1	originalRecord
6		controlfield
7	42	#text
8		tag
9	16	1
10	7	2
11	5	3
12	3	4
13	3	5
14	2	provider
15	9	id
16	10	name
17	2	_id
18		datafield
19		subfield
20		code
21		ind1
22		ind2
23	3	6
24	3	7
25	3	8
26	2	9
27	2	10
28	2	11
29	2	12
30	2	13
31	2	14
32	2	15
33	1	16
34		leader
35	1	object
36	1	aggregatedCHO
37	1	ingestDate
38	44	type
39	1	ingestionSequence
40	1	isShownAt
41	1	sourceResource
42	3	publisher
43	1	description
44		language
45		iso639_3
46	6	title
47	1	format
48	1	rights
49		contributor
50		creator
51	2	extent
52	1	spatial
53		country
54	1	date
55	1	begin
56	1	end
57	1	displayDate
58		specType
59	4	identifier
60	2	subject
61	1	ingestType
62	1	score
63	1	hasType
64	3	collection
65	1	relation
66	1	stateLocatedIn
67	1	physicalDescription
68	1	tmp_high_res_link
69	1	tmp_image_id
70	2	usage
71	2	namePart
72	7	valueURI
73	8	authority
74	8	displayLabel
75	4	titleInfo
76	1	supplied
77	1	tmp_rights_statement
78	3	relatedItem
79	1	note
80	1	17
81	1	18
82	1	tmp_item_link
83	1	originInfo
84	1	dateIssued
85	1	place
86	1	placeTerm
87	1	location
88	2	shelfLocator
89	1	physicalLocation
90	1	version
91	1	genre
92	1	rightsStatementURI
93	1	schemaLocation
94	1	typeOfResource
95	2	topic
96	2	geographic
97		stringValue
98		keyDate
99		encoding

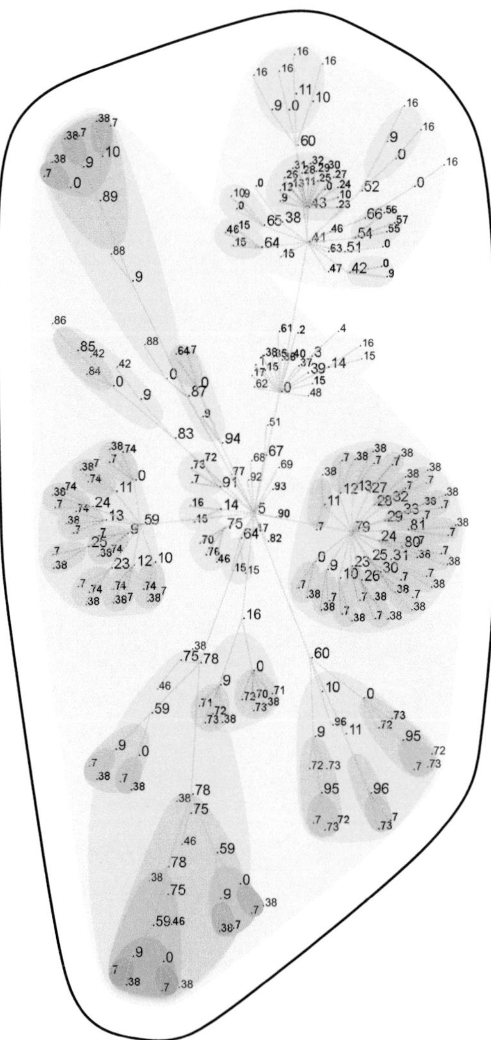

0	15	0
1	1	context
2	1	dataProvider
3		admin
4		object_status
5	1	originalRecord
6	1	controlfield
7	29	#text
8	18	tag
9	7	1
10	5	2
11	3	3
12	3	4
13	3	5
14	2	provider
15	5	id
16	3	name
17	2	_id
18	1	datafield
19	12	subfield
20	23	code
21	12	ind1
22	12	ind2
23	2	6
24	2	7
25	2	8
26	1	9
27	1	10
28	1	11
29		12
30		13
31		14
32		15
33		16
34	1	leader
35		object
36	1	aggregatedCHO
37	1	ingestDate
38	2	type
39	1	ingestionSequence
40	1	isShownAt
41	1	sourceResource
42		publisher
43		description
44	1	language
45	1	iso639_3
46	1	title
47	1	format
48	1	rights
49		contributor
50	1	creator
51	1	extent
52		spatial
53		country
54	1	date
55	1	begin
56	1	end
57	1	displayDate
58	1	specType
59	1	identifier
60		subject
61	1	ingestType
62	1	score
63		hasType
64		collection
65		relation
66		stateLocatedIn
67		physicalDescription
68		tmp_high_res_link
69		tmp_image_id
70		usage
71		namePart
72		valueURI
73		authority
74		displayLabel
75		titleInfo
76		supplied
77		tmp_rights_statement
78		relatedItem
79		note
80		17
81		18
82		tmp_item_link
83		originInfo
84		dateIssued
85		place
86		placeTerm
87		location
88		shelfLocator
89		physicalLocation
90		version
91		genre
92		rightsStatementURI
93		schemaLocation
94		typeOfResource
95		topic
96		geographic
97		stringValue
98		keyDate
99		encoding

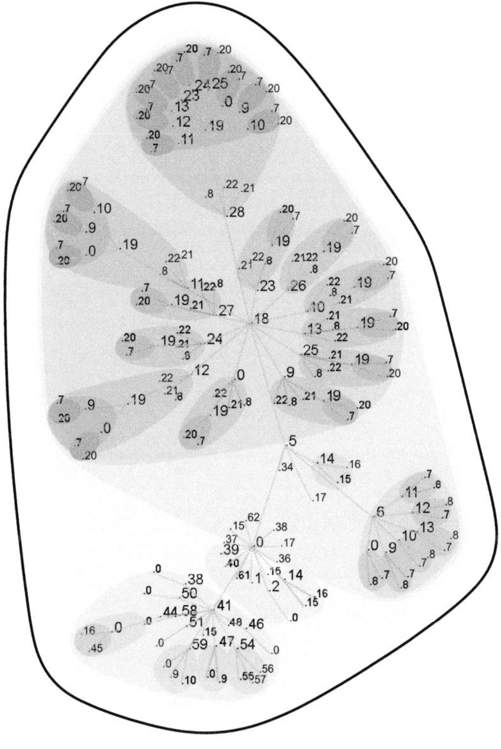

0	10	0
1	1	context
2	1	dataProvider
3		admin
4		object_status
5	1	originalRecord
6		controlfield
7		#text
8		tag
9	1	1
10		2
11		3
12		4
13		5
14	1	provider
15	4	id
16	3	name
17	1	_id
18		datafield
19		subfield
20		code
21		ind1
22		ind2
23		6
24		7
25		8
26		9
27		10
28		11
29		12
30		13
31		14
32		15
33		16
34		leader
35	1	object
36	1	aggregatedCHO
37	1	ingestDate
38	2	type
39		ingestionSequence
40	1	isShownAt
41	1	sourceResource
42		publisher
43	1	description
44		language
45		iso639_3
46	2	title
47	1	format
48	1	rights
49		contributor
50		creator
51		extent
52	1	spatial
53		country
54	1	date
55	1	begin
56	1	end
57	1	displayDate
58		specType
59		identifier
60	1	subject
61	1	ingestType
62	1	score
63		hasType
64	1	collection
65	1	relation
66		stateLocatedIn
67		physicalDescription
68		tmp_high_res_link
69		tmp_image_id
70		usage
71		namePart
72		valueURI
73		authority
74		displayLabel
75		titleInfo
76		supplied
77		tmp_rights_statement
78		relatedItem
79		note
80		17
81		18
82		tmp_item_link
83		originInfo
84		dateIssued
85		place
86		placeTerm
87		location
88		shelfLocator
89		physicalLocation
90		version
91		genre
92		rightsStatementURI
93		schemaLocation
94		typeOfResource
95		topic
96		geographic
97	1	stringValue
98		keyDate
99		encoding

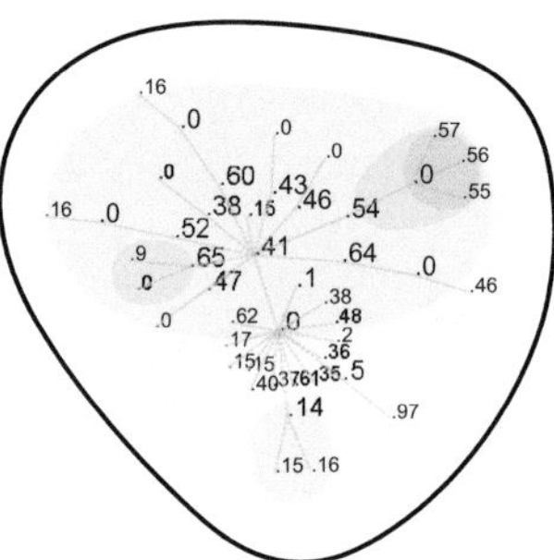

0	8	0
1	1	context
2	1	dataProvider
3		admin
4		object_status
5	1	originalRecord
6		controlfield
7		#text
8		tag
9		1
10		2
11		3
12		4
13		5
14	1	provider
15	4	id
16	2	name
17	1	_id
18		datafield
19		subfield
20		code
21		ind1
22		ind2
23		6
24		7
25		8
26		9
27		10
28		11
29		12
30		13
31		14
32		15
33		16
34		leader
35	1	object
36	1	aggregatedCHO
37	1	ingestDate
38	2	type
39		ingestionSequence
40	1	isShownAt
41	1	sourceResource
42		publisher
43	1	description
44		language
45		iso639_3
46	2	title
47	1	format
48	1	rights
49		contributor
50		creator
51		extent
52		spatial
53		country
54		date
55		begin
56		end
57		displayDate
58		specType
59		identifier
60	1	subject
61	1	ingestType
62	1	score
63		hasType
64	1	collection
65	1	relation
66		stateLocatedIn
67		physicalDescription
68		tmp_high_res_link
69		tmp_image_id
70		usage
71		namePart
72		valueURI
73		authority
74		displayLabel
75		titleInfo
76		supplied
77		tmp_rights_statement
78		relatedItem
79		note
80		17
81		18
82		tmp_item_link
83		originInfo
84		dateIssued
85		place
86		placeTerm
87		location
88		shelfLocator
89		physicalLocation
90		version
91		genre
92		rightsStatementURI
93		schemaLocation
94		typeOfResource
95		topic
96		geographic
97	1	stringValue
98		keyDate
99		encoding

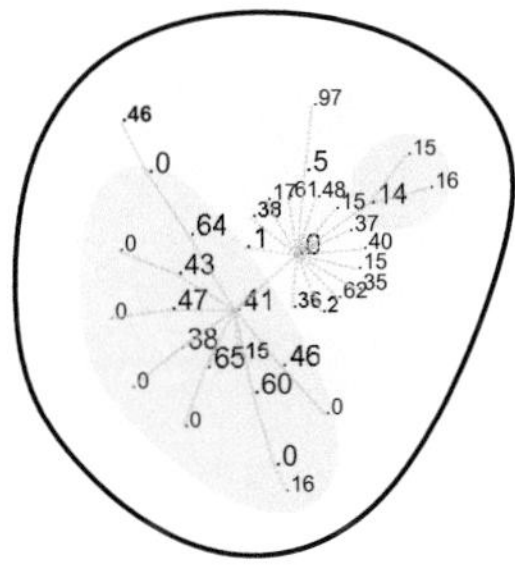

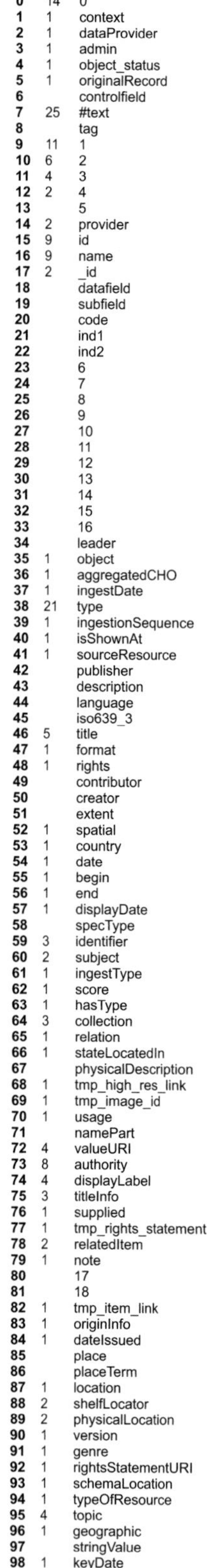

0	14	0
1	1	context
2	1	dataProvider
3	1	admin
4	1	object_status
5	1	originalRecord
6		controlfield
7	25	#text
8		tag
9	11	1
10	6	2
11	4	3
12	2	4
13		5
14	2	provider
15	9	id
16	9	name
17	2	_id
18		datafield
19		subfield
20		code
21		ind1
22		ind2
23		6
24		7
25		8
26		9
27		10
28		11
29		12
30		13
31		14
32		15
33		16
34		leader
35	1	object
36	1	aggregatedCHO
37	1	ingestDate
38	21	type
39	1	ingestionSequence
40	1	isShownAt
41	1	sourceResource
42		publisher
43		description
44		language
45		iso639_3
46	5	title
47	1	format
48	1	rights
49		contributor
50		creator
51		extent
52	1	spatial
53	1	country
54	1	date
55	1	begin
56	1	end
57	1	displayDate
58		specType
59	3	identifier
60	2	subject
61	1	ingestType
62	1	score
63	1	hasType
64	3	collection
65	1	relation
66	1	stateLocatedIn
67		physicalDescription
68	1	tmp_high_res_link
69	1	tmp_image_id
70	1	usage
71		namePart
72	4	valueURI
73	8	authority
74	4	displayLabel
75	3	titleInfo
76	1	supplied
77	1	tmp_rights_statement
78	2	relatedItem
79	1	note
80		17
81		18
82	1	tmp_item_link
83	1	originInfo
84	1	dateIssued
85		place
86		placeTerm
87	1	location
88	2	shelfLocator
89	2	physicalLocation
90	1	version
91	1	genre
92	1	rightsStatementURI
93	1	schemaLocation
94	1	typeOfResource
95	4	topic
96	1	geographic
97		stringValue
98	1	keyDate
99	1	encoding

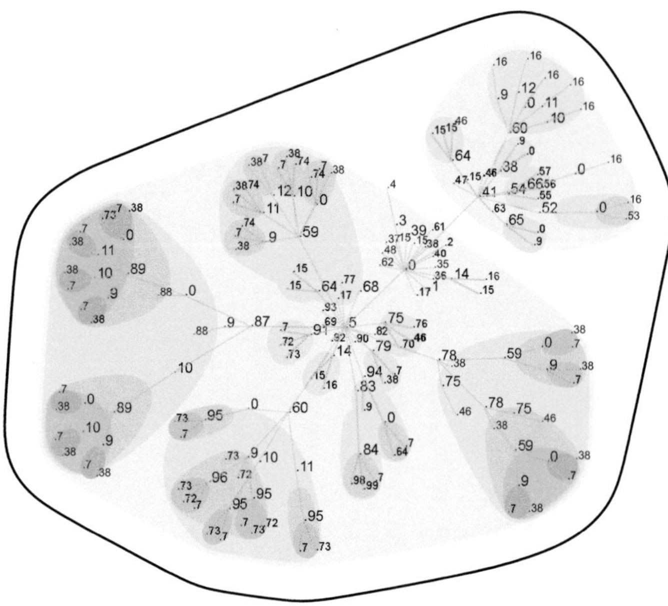

> There is a growing mountain of research. But there is increased evidence that we are being bogged down today as specialization extends. The investigator is staggered by the findings and conclusions of thousands of other workers—conclusions which he cannot find time to grasp, much less to remember, as they appear.

This passage offers a familiar depiction of a contemporary problem: an individual, working alone, is overwhelmed by sources beyond the scope of their expertise, collected by other people, in unknown times and places. This scene was described by the founder of the US National Science Foundation, Vannevar Bush, in 1945, thirty years before the invention of the personal computer. It is excerpted from his popular essay "As We May Think," published in the *Atlantic* magazine, which has since become a landmark in the history of computing.[33]

Bush's proposed solution to the "mountain of research," the memex (short for "memory expansion"), captures an early desire for a space in which a single investigator might navigate and connect data from disparate origins. He depicted the memex as an ordinary World War II era desk, outfitted with a pair of projection screens, a keyboard, an array of buttons and levers, and a memory store of unprecedented size for the time—all neatly concealed under the desk's working surface. Harnessing a combination of dry photography and microfilm, the latest technologies of the postwar era, Bush believed that the machine might be capable of storing millions of independently addressed records: books, newspapers, photographs, and correspondence. The memex, writes Bush, would "instantly bring files and material on any subject to the operator's fingertips."[34]

The idealized setting for data heralded by Bush's machine—individualized, isolated, and unsurprisingly placeless—still speaks to the creators of today's data infrastructures. Though never built, the memex articulated an emergent desire to liberate data from their attachments to institutionalized settings, which Bush argues are only obstructions to knowledge discovery:

> Our ineptitude in getting at the record is largely caused by the artificiality of systems of indexing. When data of any sort are placed in storage, they are filed alphabetically or numerically, and information is found (when it is) by tracing it down from subclass to subclass. It can be in only one place, unless duplicates are used; one has to have rules as to which path will locate it, and the rules are cumbersome. Having found one item, moreover, one has to emerge from the system and reenter on a new path.[35]

Bush imagined that the memex could seamlessly extend human thinking—a function evoked in the title of his essay. He wanted a machine that, as Orit Halpern explains, would "break the taxonomic and stable structure of the archive and would work 'as we may think,' by creating rhizomatic linkages and nonlinear associations between different pieces of information."[36] To enable this nascent desire, Bush introduced the notion of *associative trails*, a precursor to what we now call *hyperlinks*, specifically meant to automate out librarians and catalogers.[37] He saw little merit in the work of "clerks," offering his associative trails as a more "natural" alternative.[38] Bush wanted the investigator to be free to encounter data in a frictionless realm representative of and yet set apart from the world.

A similar ambition guides the development of contemporary data infrastructures, designed to collect, maintain, and distribute data across network technologies.[39] Indeed, contemporary data infrastructures act as Bush envisioned, but only in a superficial sense. They help scientists—but also educators, professionals, and an increasingly broad public—manage streams of data that would otherwise overwhelm an individual. Yet Bush did not predict some of the most important social changes in practices with data—changes that would make data simultaneously smaller and bigger than he could have imagined.

First, knowledge practices have rapidly diverged, leading to a variety of data cultures, each of which manages data in its own way. Second, data are widely distributed. The notion of a personal database has been replaced by a fascination with the potential of the web as a platform for access to data from almost anywhere. Third, data have become big business. As in the example of Zillow explored in chapter 5, the potential surplus value of data has stimulated aggregation, thereby, severing crucial ties between data and the local institutions in which they are made.

Learning to look for the local in data can help us see data infrastructures as composites; they juxtapose classifications, schemata, constraints, errors, absences, and rituals from diverse data sources. Data infrastructures would do well to acknowledge the history of collecting practices, for some day those same infrastructures will be historical relics too, and their choices will constrain future generations.

Seeing that all data are local means dismantling the image of data work manifest in the memex, a machine envisaged for a single investigator, a scientist (and man), sitting at a command-and-control center with access to the vast store of human knowledge. Instead, we must acknowledge that working with data is social; it requires continual communication and care.

I see this as an ethical shift from the model of the master sitting at his desk of power. We might yet move to a condition in which more intimate relationships with data and their subjects are widely fostered as an ethical responsibility. This is not some radical speculative future. It is demonstrated everyday by librarians, archivists, arboretum workers, and even some realtors.

One final caution about data infrastructures: simply using them can reinforce the social significance of the large, well-supported organizations that create and contribute to them. Thus we also need counterdata or even antidata: tactical representations that challenge the dominant uses of data to secure cultural hegemony, reinforce state power, or simply increase profit. Counterdata might encode alternative perspectives to the cultural histories of the DPLA. Although counterdata have their own limits, they are necessary to counteract dominant representations of the past and ways of imagining the future; an alternative to the memex is long overdue.[40]

CONCLUSION

In the *New York Review of Books*, Robert Darton sought to characterize the early efforts of the DPLA as an initiative that is at once universal and thoroughly American: "A library without walls that will extend everywhere and contain nearly everything available in the walled-in repositories of human culture. ... E pluribus unum! Jefferson would have loved it."[41]

The tension expressed in this quote, between the all-encompassing ambitions of present-day data infrastructures and their local origins, attachments, and values—in this case, expressly US values—is a common refrain throughout this book. Darton portrays the DPLA as a "meta-mega-macro library," a collection that could, in principle, contain everything. It is an old ambition that has only recently begun to seem like a possibility. But the name of the initiative (the Digital Public Library of *America*) alongside Darton's evocation of the slave-owning "Sage of Monticello" should raise questions about the actual scope of the DPLA. Whose America does it represent? Moreover, cultural repositories are not simply "walled-in." Walls are cultural constructions in their own right. They are a means of making a social order durable and hence important for understanding everything contained therein. Similarly, the unedited collections data contributed to the DPLA are the structural elements of its cultural histories, registering the norms that unite and separate diverse cultures of collecting within the United States.

Today there are a proliferation of initiatives that assemble data infrastructures from local, distributed sources, as is illustrated by the DPLA. Each data infrastructure promises to reveal new patterns across previously independent data sets. Yet these initiatives are not all the same in their motivations and goals. The DPLA is an example from the nonprofit world. Created with a mission "to educate, inform, and empower everyone in current and future generations" (www.dp.la), the DPLA is supported by a combination of private foundations, individual philanthropists, and US federal research agencies. Meanwhile, academia and industry have their own models. The next two cases in the book are also data infrastructures, but with different ends and means. NewsScape, an academic model supported entirely by public funds, is used for research in disciplines ranging from communication to computer science. Zillow is a model from the world of industry: a for-profit platform motivated by the potential for surplus value gained through the aggregation of data.

Despite their differences, the DPLA, NewsScape, and Zillow are unified by a consistent motivation: to assemble collections that are seemingly complete. Their creators aspire to build comprehensive perspectives on the world, offering vistas across all the books, all the news, or all the real estate opportunities. We might think of data infrastructures as an instance of what David Nye calls the "technological sublime," for they test the limits of human perception and imagination.[42] But as these cases reveal, data infrastructures only appear to be comprehensive. As I demonstrate, they have noteworthy limitations. There is an emergent need for alternatives to the universalizing discourses that surround data infrastructures, infusing them with a sense of truth and objectivity.[43]

I propose that working effectively and ethically with data infrastructures means seeing them through a comparative lens—by acknowledging the whole as well as its local, heterogeneous parts. Local readings of data infrastructures, informed by interviews with those who make and use data, can reveal the variety of local ties that they harbor in classifications, schemata, constraints, errors, absences, and rituals. Seeing data infrastructures as assembled from local conditions opens up new opportunities and obligations for scholarship as well as pedagogy and practice. Yet we shouldn't romanticize local ties; as some examples illustrate, they can be lacking in sophistication or even be discriminatory.

In the next two chapters, I deal with external though salient dimensions of data infrastructures: the algorithms that activate them, and the interfaces that recontextualize them. In chapter 4, I use NewsScape to uncover the deep historical and material entanglements between data and algorithms. In chapter 5, I explain, via Zillow, how we can develop critical perspectives on the mechanisms—visual, discursive, and algorithmic—by which interfaces make data actionable.

4

NEWSWORTHY ALGORITHMS

```
2016–11–09_0700_US_CNN_Election_Night_in_America.txt,22913eee-
a64a-11e6-bdad-089e01ba0338,election,11/9/16 7:09, "THE LEADER
OF THE DEMOCRATIC PARTY, AND FOR HER NOT TO WALK –>> IS IT TOO
LATE?>> IS IT TOO LATE?>> YEAH.>> I THINK AS WE'VE JUST SEEN,
THIS IS AN ELECTION THAT IS DEEPLY DIVIDED THE COUNTRY. THE
POLLING—WE LOOK AT—WE'LL PROBABLY GET THE POLLING OVER NIGHT,
BUT THE POLLING GOING IN WAS 90% OF THE HILLARY CLINTON VOTERS
","http://www.sscnet.ucla.edu/tna/edge/video,22913eee-a64a-11e6-
bdad-089e01ba0338,549000"
```

Source: NewsScape Archive[1]

ALL NEWS IS LOCAL

Saturating the ensuing pages (figure 4.1) is a list of words, each one extracted from news coverage of the 2016 US presidential election: a contest in which Donald Trump upset widespread predictions that Hillary Clinton would become the first woman commander in chief. The list begins neutrally enough, with the election characterized as simply *different*, *general*, and *important*. But the discourse soon progresses to *pivotal*, *polarizing*, and *bizarre*. At its midpoint, the election campaign is *compromised*, *looming*, and *zombie*. It ends with proclamations of *magic* and *awful* but *secure*.

All these words immediately precede the term *election* in news coverage from a period starting 418 days before the vote was held, during the first candidate debate on September 16, 2015. The list ends on election day, November 8, 2016. These words—all descriptors for the election, of one sort or another—appear in the order they were used by one of three news outlets: Cable News Network (CNN), *Wall Street Journal* (*WSJ*), and Breitbart News Network (BNN). Each word appears only once in the list, at the moment of its first use (since September 16, 2015). Color indicates the news outlet that first used it: black for CNN, blue for the *WSJ*, and red for BNN. The numbers that precede each word indicate the days left until the election. Those that follow, also colored black, blue, and red, indicate how many times the term was subsequently invoked by each news outlet.

4.1 (following pages)

The NewsSpeak algorithm returns a list of words, which precede the term *election*, in news coverage from multiple sources. Image by the author and Peter Polack.

Days until election		CNN	BNN	WSJ
418	a	18		
	an	1958	24	126
	different	30		
	general	6248	9	202
	important	161		2
	last	219	1	3
	losing	10		
	presidential	1701	17	275
	recall	12		1
	school	3		
	the	8832	62	345
	these	115	1	
	this	6280	24	85
	unusual	29		2
417	-	945	7	254
	2016	151	1	53
	buy	81		1
	come	46	1	
	competitive	14		1
	faso	1		
	first	74		2
	for	144	1	5
	huge	25		
	next	101	1	2
	six	42		2
	statewide	20		
	t	3		1
	that	260	3	9
	three	21		2
	welcome	2		
	winning	44		
416	.	567	5	68
	committing	1		
	federal	112		40
414	four	13		2
	his	114	2	22
	national	153	4	20
	nec	1		
413	early	9	2	8
	fair	87		5
	have	26		
	republican	21		2
	resounding			1
	saying	5		
412	?	12		
	every	205		5
	from	170	1	2
	greek			1
	rigorous	2		
	win	108	1	5
	your	20		
411	'	6		3
	2012	127	1	28
	to	354	2	3
410	1896			13
	2014	3		6
	entire	121		
	her	21	1	
	particularly	1		
	primary	151	10	23
409	entering	2		
	midterm	52		17
408	s	237	16	62
406	assad			1
	buying	24		1
	democratic	35		5
	landslide	12	1	3
405	full	2		
	leadership	14		4
	previous	69	3	1
404	and	207	1	13
	one	50		3
	separatist	1		
	until	418		2
403	more	16		1
402	coming	13	1	14
	local	57	5	7
	many	41		
	of	380	11	37
	own	47		2
401	2009		1	2
	hold	8		2
	legislative		3	5
400	so	6		
	way	8		
399	critical	21		1
	major	6	1	2
	robust	1		
	speaker	4		1
398	ballot	3	1	2
397	2000	43		1
	changed	24		
	close	165		3
	my	20		
	on	1081	8	44
	stolen	21		
	where	3		1
396	2008	37		6
	29			1
	big	39		2
	called	1	2	

Days until election		CNN	BNN	WSJ
	class	4		
	floor			1
	internal			2
	obama	15		2
	out	9		
	seeking	4		
	short	1		
	special	32		4
	those	39	3	2
	with	37	4	5
394	parliamentarian	1		
	parliamentary	25	4	17
	upcoming	55	2	
393	lexl	1		
392	1999			1
	contentious	8		2
	delegates	1		
	fall	52		5
	mobilization	9		
	state	87	3	9
	swung			1
	traditional	6		
	two	116	2	2
	year	14		1
391	any	54		2
	change	101		
	in	227	9	13
	larger	3		
390	breaks	1		
	few	25		
	free	22		2
	growth			1
	president	7		
	recent	45	1	7
389	actual	35		
	fascinating	5		
	flower	2		
	historic	45		2
	house	6		1
	june		6	1
	potential	2		1
388	19		1	
	into	74	1	
	lose	8		
	prior	8		3
	whether	1		2
387	another	44	2	1
	congressional	6	1	6
386	new	53		3
385	canadian	4	1	
	future	11	1	1
	whole	104	1	1
384	current	17	1	1
	director			2
	past	92		7
	sixth	1		
	staggered			1
383	be	18		
	conference	2		
	particular	37		
382	1998			1
	at	19		2
	campaign	4		
	he	5		
	positive	7		
	successful	8		2
381	--	17		
	jgeneral	3		
378	breaking	5	2	
	broke		1	
377	speakers	1		
	volatile	11		
376	1994	4		1
	is	70	1	
	wild	8		
375	crazy	19		
	easy	4		
	senate	17		
374	book	1		
	endless	22		
	hard	3		
	some	16		1
	think	14		1
372	closest	11		
	since	2		
	snap	4	1	
371	actually	3	1	
	because	16		1
	been	15		
	call	5		
	clean			2
	cliffhanger	1		
	council	4		1
	later		1	
	secretary		1	
	tendless	1		
370	country	9		
	each	22		1
	ele	22		
	happy	8		
	independent			2

Days until election		CNN	BNN	WSJ
	jrl	1		
	newsletterearly		1	
	october			3
	prnl	2		
	unexpected	1		2
	yes	2		
369	about	83		2
	hope	3		
	interesting	18		
	partisan	2		
	polarized	4		2
	sweep	2		
	taiwanese		2	1
368	addition			2
367	anymore	1		
	month	1		
365	by	46	2	1
	freest	1		
	myanmar	2		
	was		1	
364	---			1
	turvy	2		
363	heated	9		3
	impact	5		
	only	22		1
	strange	44		1
	weird	11		
362	6	6		
	having	7		
	our	367	2	2
	said	11		
	their	42		6
	track	2		
360	contested	26		3
	runaway			1
358	french	7		1
	yep	2		
357	political	29		
352	facing	3		
351	runoff	10	1	
350	regional	8	2	3
	twitter	3		
349	constitution	1		
347	british	11		1
	through	55		3
346	focused	2		
	post	21		
344	get	1		
343	clinton	8		
	official	3	1	
	private	1	3	
342	time	6		1
341	atypical	5		
	union			2
339	policy	2		
338	after	113		7
335	electione	1		
	second	4		
334	basing	2		
	e	9		
	liberty	1		
333	rag	1		
	unpredictable	13		
332	drive	1		
	electoral	1		
	serious	7		
331	november	118		38
327	before	394	2	19
	polls	2		
326	cover	3		
	security	7		
	see	2		
324	base	9		
	or	15		4
323	fun	8		
322	may			1
321	decide	4		2
	lost	16		1
320	card	1		
	couple	16		
	inconclusive			2
	mock	11		
	other	33		2
	outsider	5		
	say	8		
	thermometer	3		
	thermostat	3		
	us	4		
319	great	9		
317	annual			2
	typical	10		
316	smooth	2		
315	several	26		1
314	dramatic			1
	liberal	1		
	loses	4		
313	comes	2		
311	mean	1		
308	8	4	1	4
	february	1	1	
307	no	17		1

Days until election		CNN	BNN	WSJ
	thought	1		
306	:	2		4
	per			1
	successive	1		2
304	during	26	2	3
	failed	4		
303	disputed	4		2
	iranian	3		2
302	as	33	1	4
	wave	21		
301	mandate	1		
300	above	1		
	el	3		
	presidency	3		
299	better	2		
	covering	5		
	day	4		
	persuasion	8		
	they	1		
	what	29		
298	won	24		
297	representation			1
	resume	2		
295	expert	2		
294	2010	4		
	although		2	
	parliament	1		
293	administration			1
	te	2		
292	crucial	8		
	extreme			5
	fought	7		4
	governor	3		
	gubernatorial	5	1	1
	municipal	4	3	1
	wins	6		
291	96	1		
	consequential	31	1	
	months			1
	moving			1
288	head	2		
287	1992	3		1
	germ	9		
	representative	2		
	skwlen	1		
	talked	1		
	tumultuous	3		3
286	american	143	1	7
	but	12		
	consecutive	7		1
	overall	3		
285	alter			1
	extraordinary	6		
	jebl	2		
	large	2		
283	"	4		11
	influence	11	3	2
	like	6		
282	between	6		2
	how	14		1
	wide	3		
281	iowa	10		
	wacky	1		
280	all	17		
	changer	1		
	dunk	1		
	normal	20		1
	sanders	4		
279	march	1		
	mayoral	2		
	ordinary	10		1
	reset	1		
	term	22		
	turnout	45		
278	20	3		1
	around	14		2
	determine	2		
	overcome			1
	professional	2		
	unfair	2		
277	over	16		
	talk	1		
	top	4		
276	prescinctpreci	2		
	sanctioned	1		
	wonderful	1		
275	tough	17		
	tumble	8		
274	disturbing	2		
	open	6		1
273	hampshire	2		
272	amazing	9		
	either	1		
271	15	4		
	cnn	118		
	covered	10		
	party	5	2	1
	processing	1		
	roll	1		
	trump	13		1
270	send	1		

Days until election		CNN	BNN	WSJ
	single	19		2
269	1824	2		
	passing	1		
	stole			1
268	raucous	2		1
	word	2		
267	clarifying	3		
	divisive	5		1
	duck	1		
	monumental	2		
	tara	1		
265	consider	1		
	manipulating	1		
	normally			1
	protracted	1		1
264	are	30		
	f	1		
	final	19		
	make	4		1
	pivotal	6		1
	real	13		
	someone	2		
263	unique	13		
262	appn	1		
	five	21		
	polarizing	4		1
	woolly	2		
260	butter	1		
	difficult	7		
	states	14	1	2
	tight	26		2
259	[			1
	predicting	1		
258	says	2		
	today	3		
257	]	1		
	08	7		
	candidate			1
	good	11		
	gop	3		1
	held	3	1	2
	holding	5		
	nn	2		
	washington	2		
256	dirty	1		
255	ll	1		
254	carolina	6		
	numbers	1		
253	ten	1		
252	1996	3		
	jen	1		
	modern	16		1
251	1972	2		
	board	4		4
	broad	1		
	generally	3		
	had	16		
	hit	3		
	stakes	3		
	when	23		
250	1828			1
	apop	1		
	rags	1		
	supercompetitive	1		
249	1960	4		2
	colorful	3		
	inevitable	2		
	sleepy	2		
	throughout	4		
	unprecedented	15		
248	app	1		
	weekend	1		
	which	6		
247	closed	3		
246	eve	1		
	funded	1		
	half	3		
245	1984	5		
	nationwide	4		1
244	county	14		
	deadlock	1		
243	direct	6		
	handles	1		
	hypothetical			1
	macro	2		
	organizing			2
	same	5		
	surreal	4		
	suze	1		
242	15th	1		
	1980			1
241	look	5		
	ohio	1		2
	possible	1		1
	run	32		
	running	2		
	vigorous	1		
240	peaceful	2		
239	1860	1		
238	know	10		
	worst	6		1

Days until election		CNN	BNN	WSJ
237	2004	5		3
	certify	1		
	closer	6		
	florida	7		
	key	5		2
	mature	1		
	tuesday	20		
236	angriest	1		
	complex			1
	gentle	2		
	hillary	2		
	toxic	4		
	volcanic	1		
235	against	5		
	beyond	3		
	explosive	1		
234	nominating	3		
233	2018	1		1
	utah	2		
232	heavy	2		
	pending	6		
	power	2		
231	beg	1		
	exciting	12		
	remember	1		
230	tree	2		
229	average			1
	bipartisan	4		
	sol	2		
	surprise			1
	western	3		
228	immediate	1		
226	bizarre	5		2
	complete	7		
	remarkable	1		
	separate	2		
	september			1
	thus	1		
225	1948	2		
	emotional	2		
224	nasty	16		
223	27	1		
	54	1		
	7	1		
	seven	8		
221	2013			2
	changing	4		
220	both	11	1	1
218	century	1		
	rifting	1		
	stronger	1		
216	bush	1		
	governors	4		
	judicial	4		
	will	2		
215	contest	5		
	cruz	3		
	formal	3		
	hear	2		
	you	2		
214	towards	13		
213	delegate	18		
	disrupted	2		
212	30	2		
210	12	4		
	colorado	2		
	most	22		
	multiple	7		
	posthumous	1		
	regular	6		
	sponsored	2		
209	11	12		
	cancelled	1		
	swept	1		
208	catastrophic	1		1
	given	3		1
	risky	2		
207	decisive	3		
206	reform	1		
205	cancel	6		
	difference	2		
204	media	2		
	social	1		
203	crooked	2		
	dreamy	1		
	preference		1	
	rigged	572	3	2
	transparent	3		
202	electi	2		
	empowered	2		
	influencing	4		
	pluralities	1		
	question	3		
201	1990	1		
	does	1		
	outsiders	5		
	voterless	6		
	voting	6		
200	newy	1		
196	individual	3		
	sign	1		

Days until election		CNN	BNN	WSJ
195	confident	2		
	course	4		
	fourth	2		
	mood	8		
	predict	1		
	ultimate	6		
194	!	1		
	biggest	5		
	genderal	1		
	likely	1		
	message	2		
	repeat			1
193	eeral	1		
	referendum	2		
192	home	4		
191	california	3		
	consequences	2		
	help	1		
188	benefits	1		
	gets	1		
	presidenti	2		
187	its	4		
186	mexican	1		
185	choice	4		
	moment	2		
	turbulent			1
184	coalition	1		
	incredible	3		
183	2006	1		
	theory	2		
182	cut	1		
	understand	1		
181	50	1		
	brutal	5		
	gap	1		
	gem	2		
	nastiest	2		
	romney	1		
180	immobilization	2		
	le	4		
	overnight	2		
	spirited	1		
179	forry	1		
	usual	11		
178	looks	2		
	puzzling	1		
	research	2		
	ugly	11		
177	conventional	6		
176	faulty	2		
175	alex	1		
	controversial	2		1
174	candidates	3		
	claims			1
	lbj	3		
173	administer	1		
	chief	6		
171	putting	5		1
170	popular	2	1	
168	2015	1		1
	corporate			1
	layup	2		
	nose	2		
	uninspiring	1		
167	fantastic	1		
	swings	3		
	violated	1		
166	elec	3		
	job	1		
165	antiestablishment	1		
	break	8		
	establishment	3		
	german	1		1
	movement	2		
164	main	1		
162	financing	1		
	oning	1		
	sometimes	1		
	sway	3		
161	8th	20		
	mcgovern	1		
160	various	3		
159	legacy	1		
	tv	1		
158	64	2		
	rare	3		
156	boxing	1		
	probably	2		
155	genera	1		
154	dakota	2		
	still	2		
153	gat	3		
	superdelegate	2		
152	flagship	1		
	n	1		
	odd	4		
151	ae	1		
	hand	1		
150	2002	1		
	surprising			1
149	1964	2		
	equal	2		
144	shall	1		
	tougher	1		
	well	2		
141	provincial	3		
140	broader	3		
	then	3		
139	based	1	2	
	outside	2		
	unconventional	6		
138	bag	1		
	released	1		
137	20916	1		
	effective	1		
136	collection	1		
	dollar	1		
	european	2		
	overwhelming	4		
	th	3		
133	versus	1		
132	2			1
	right	8		
	shl	2		
131	04	1		
	den	3		
130	1940	1		
127	there	1		
125	face	1		
122	collects			1
120	australian	1		
119	following	2		
116	point	2		
115	near	2		1
113	indiana	1		
111	truth	2		
110	bowl	1		
	harder	2		
109	%	4		
	accepting	1		
	dprl	1		
107	beautiful	3		
	personality	4		
	unending			1
106	insurgent	1		
104	straight	2		
103	quo	1		
102	wrong	3		
101	countries	2		
	definitional	1		
	funding	2		
	goat	2		
	negative	2		
	up	6		
	using	2		
99	affects	1		
	were	8		
98	quote	1		
	quoting	1		
97	politics	5		
96	global	4		
	uncertain			1
95	costs	2		
	honest	5		
	reflective	2		
94	crossroads	5		
	questionable	2		
	umpteenth	1		
93	believe	1		1
	compromised	8		
	perhaps	2		
	respective	1		
92	historical	3		
90	13	1		
	emptive	3		
	gore	3		
	investigating	1		
	profile	1		
	steal	7		
	uplifting	2		
89	hack	2		
	lebanese	3		
	phony	6		
88	specifically	1		
86	become	2		
	bombshell	2		
	do	4		
	legit	1		
	legitimate	3		
	recruit	2		
	shell	2		
	tb	2		
	tur	2		
84	observe	1		
	rig	5		
83	despite			1
	i	3		
	mega	1		
	rae	1		
	third	4		
81	contrast	1		
	whacky	1		
79	finally	1		
	virginia	1		
76	rigging	9		
73	toward	3		
72	election	2		
	white	2		
71	winnable	1		
70	1988	1		
	monitor	2	1	
	viable	3		
69	arizona			2
	everything	1		
	falsify	1		
	illinois			1
	targeting	1		
	than	4		
68	characterized			1
	polish			1
67	aggressive	3		
	apples	1		
65	off	2		
	otherwise	2		
64	dispiriting	2		
	sad	11		
63	2011		2	
	foush	1		
	once	1		
	till	6		
62	trust	2		
61	ethe	2		
	life	1		
	yial	1		
58	analyze	2		
	looming	1		
	nine	2		
56	facts	1		
	ninth	2		
55	approach	6		1
	driven	2		
	even	5		
	gear	1		
54	ebb	1		
	legal	1		
	not	7		
	russian	8		
53	debates	3		
	doing	2		
	down	4		
	messy	4		
52	button	1		
	clear	40		
51	russians	1		
50	winnings	2		
48	com		1	
	performance		1	
46	making	1		2
42	certain	2		
	ddemocratic	1		
	heat	1		
	inspirational	1		
	tie	4		
41	chris	1		
	egg	1		
	straightforward	1		
40	disrupt	3		
	eat	1		
	strangest	4		
	swing	3		
	unorthodox	3		1
39	attempted	2		
	it	2		
	military			1
	tied	2		
	while	2		
38	grade	1		
37	bad	1		
	modify	1		
	slinging	1		
36	again	1		
35	true	1		
34	latest	5		
	vicious	1		
33	swan	1		
32	hoping	1		
30	31	1		
	insane	3		
	premier	1		
29	whose	6		
27	computerized	1		
	dece	1		
	fraudulent	3		
	interest	1		
	matters	1		
	officially	2		
	style	1		
	thin	1		
	though	1		
	zombie	1		
26	discouraging	2		
	downballot	2		
	why	7		

Days until election		CNN	BNN	WSJ
25	rough	1		
	sick	1		
24	apparent	1		
23	2020	1		1
	attacked	1		
	scandal	3		
22	1982	5		
	dismal	2		
	divided	18		
	example	1		
	hijacked	2		
	horrible	1		
	prosecute	2		
	prosecuted	4		
21	51	3		
	bitter	2		
	discredited	2		
	fraud	2		
	furniture	1		
	love	6		
	luck	1		
	questioning	2		
	scale	3		
	study	3		
	swayed	3		
	thinking	2		
	very	1		
	widespread	5		
20	city	3		
	conspiracy	1		
	feel	1		
	fix	2		
	fixed	3		
	govern	3		
	mind	3		
	observing	1		
	studied	4		
	traumatic	1		
	trig	2		
19	7th		2	
	if	1		
	oversee	3		
	ricked	1		
	ringed	2		
	selling	1		
	smaller	2		
	support	2		
	tabloid	1		
18	accept	6		1
	al	3		
	bogus	2		
	corrupt	2		
	decided	2		
	dozen	1		
	elect	4		
	finale	2		
	finish	2		
	orderly	2		
	rampant	1		
	thing	1		
	voted	1		
17	bruising	1		
	disputes	1		
	extended	1		
	huh	1		
	knuckles	1		
	monitoring	4		
	precise	1		
	rankerous	2		
	stealing	1		
16	anchors	2		
	defeat	2		
	nowrom	1		
	waive	2		
15	elhe	1		
	former	2		
	ged	2		
	got	2		
	magic	1		
	stressedhis	2		
14	americans	3		
	carroll	1		
	didvicive	1		
	speakership	1		
	statement	1		
	wat	1		
13	biter	2		
	celebration	2		
	chad	2		
	comparable	1		
	conduct	2		
	demographic	3		
	ending	1		
	fm	1		
	shortened	1		
	under		1	
12	club	5		
	gaffe	1		
	terrible	1		
	tremendous	1		
11	4	1		
	40			1
	bunker	1		
	chaotic	1		1
	draining	1		
	government			1
	international			2
	reagan			1
	skip	1		
	updated	1		
	warning	1		
10	david	1		
	dented	1		
	spending	6		
	undermine	1		
9	impending	1		
	I	1		
	predictable	1		
	sheriff	2		
7	bhfr	1		
	cannibalize	1		
	controlled		2	
	ecome	1		
	electronic		2	
	suppressed	1		
6	baked	1		
	citing	1		
	craziest	1		
	dull	1		
	electio	1		
	intense	1		
	now	1		
	princeton	1		
5	debate	2		
	defining	3		
	meanwhile	1		
	offee	1		
	plot	3		
	populous	1		
	thoo	1		
	voters	4		
	yeah	1		
4	america	4		
	bee	1		
	lekd	1		
	night	1		
	pg	1		
3	inhis	1		
	late	1		
	need	1		
	planning	1		
	requiring	2		
	watch	2		
	work	1		
2	able	2		
	bring	1		
	catch			1
	daye	2		
	foreign	2		
	healing	12		
	just	3		
	lag			2
	leak	1		
	live	2		
	shape	1		
	subsequent	2		
	tense	1		
	them	1		
	wildest	2		
1	1800	1		
	conscience	2		
	epic	1		
	similar	1		
	watershed	2		
0	1932			1
	being	1		
	criticizing	2		
	defines	1		
	destabilize	1		
	ensuring			1
	here	1		
	intermable	2		
	keeping	2		
	long	3		
	memorable	2		
	use	1		
	watching	1		1
	2e679ed	1		
	awful	1		
	especially	3		
	hopeful	1		
	justice	2		
	means	1		
	poll	1		
	preliminary	1		
	roughest			1
	secure	1		

For example, the fifth word used to describe the election is *important*. It comes from this statement made on CNN in the first few moments of a debate on September 16, 2015: "This is an *important* election with an enormous number of challenges facing the American people and the first four questions are about Donald Trump." *Important* was eventually used directly before *election* 161 times by CNN and twice by the *WSJ*. Based on this list, BNN does not seem to have used the word *important* to characterize the election at all.

Assembled together on these pages using a simple algorithm (or computer program) that I created, called NewsSpeak, these words offer a rule-based reading of election news as it might be revealed through data.[2] NewsSpeak is not intended to offer a comprehensive view of election coverage. Nor is it meant to produce a political analysis of the news during the eventful period that it covers. The algorithm is much more mundane: it calls attention to important processes that are too often taken for granted as routine in contemporary practices of computing applied to text. NewsSpeak is designed to be an encounter with the locality of "natural language" data.

The bulk of the words listed by NewsSpeak, specifically those extracted from CNN coverage, were obtained from NewsScape, an online television news archive based at the University of California at Los Angeles that has assembled more than four hundred thousand broadcasts from around the world, some of which date back to the US Watergate scandal in the early 1970s.[3] The data excerpt at the beginning of this chapter provides an example of that data in its original form, as it was digitized from CNN's early morning coverage on November 9, 2016, the day after the election. The date and station identification are listed first, followed by text derived from closed captioning. The final string is an internet link that leads directly to an excerpted online video of the coverage.

NewsScape is one of few such large-scale archives that seek to assemble comprehensive collections of the news as data.[4] But NewsScape doesn't contain newspaper or online journalism. The two other sources used in NewSpeak, the *WSJ* and BNN, are scraped from ProQuest and the BNN site, respectively, using yet another algorithm,

4.2

CNN video snapshot from November 9, 2016, attained using the Internet Archive TV News Archive.

4.3

Wall Street Journal front page from November 9, 2016, archived by the *New York Times*. Victor, "Trump's Victory, on Front Pages Worldwide."

THE WALL STREET JOURNAL.

DOW JONES | News Corp ********* WEDNESDAY, NOVEMBER 9, 2016 - VOL. CCLXVIII NO. 111 WSJ.com ★★★★ $3.00

DJIA 18332.74 ▲ 73.14 0.4% NASDAQ 5193.49 ▲ 0.5% STOXX 600 334.91 ▲ 0.3% 10-YR. TREAS. ▼ 12/32, yield 1.867% OIL $44.98 ▲ $0.09 GOLD $1,273.40 ▼ $4.90 EURO $1.1025 YEN 105.17

PRESIDENT TRUMP

POPULIST SURGE LIFTS REPUBLICAN TO UPSET

Clinton lost in key battleground states; GOP keeps Senate and House

What's News

Business & Finance

Electoral gains by Trump on Tuesday evening spurred a sharp decline in U.S. stock futures and a broad flight to safety around the globe. Government bond yields slid. A1, C1

◆ **U.S. businesses** will face uncertainties with a Trump win in the election. A3

◆ **Walgreen sued** Theranos, alleging that its former laboratory-testing partner breached a contract between the two companies. B1

◆ **A Chinese aluminum firm** is evading U.S. trade sanctions, the government said, as it tries to rein in a flood of cheap metal imports. B1

◆ **Alphabet ousted** two managers on its drone-delivery project amid infighting on its team. B1

◆ **Amazon could be** in the crosshairs of Europe's taxman, and the stakes for the online retailer are high. B7

◆ **CVS warned** that it could lose 40 million prescriptions next year as Walgreens deals shut out CVS pharmacies. B2

◆ **Valeant cut** its annual forecast again, sending shares tumbling 22%. B2

◆ **Icahn more** than doubled his Hertz stake, the same day the rental-car firm's stock sank 22.5%. B3

World-Wide

◆ **Trump won** the 2016 election, becoming the nation's 45th president by defeating Clinton in one of the biggest upsets in U.S. political history. Defying polls and political pundits, the Republican triumphed over his Democratic rival by running up extraordinary vote totals in rural areas and working-class counties across the U.S. where his message of economic populism resonated. A1

◆ **Republicans captured** a series of hard-fought Senate races Tuesday night, allowing the party to retain the chamber's majority and solidify complete GOP control of Congress. A4

◆ **Democrats registered** only modest gains in the House, falling short of their most ambitious plans to put a significant dent in the Republican majority. A4

◆ **Turkish guards** abused political prisoners in the wake of the July coup attempt, eight of the prisoners alleged in a letter. A15

◆ **Iraq began** to identify bodies in a mass grave in a village near Mosul, from which Islamic State reportedly forced out residents. A16

◆ **Hungarian lawmakers** rejected the prime minister's proposed ban on future EU resettlement of migrants in the country. A16

CONTENTS
Arts in Review ... D5
Business News ... B2-3,6-7
Crossword ... B7
Election 2016 ... A2-13
Global Finance ... C3
Heard on Street ... C14
In the Markets ... C4
Opinion ... A17-19
Property Report ... C8-10
Sports ... D6
U.S. News ... A14
Weather ... B7
World News ... A15-16

President-elect Donald Trump addresses supporters in Manhattan early Wednesday after his stunning upset of Hillary Clinton.

Markets Jittery on Early Results

By Ira Iosebashvili, Min Zeng and Chelsey Dulaney

Electoral gains by Republican presidential nominee Donald Trump on Tuesday evening spurred a sharp decline in U.S. stock futures and a broad flight to safety around the globe, underscoring trepidation among investors that the New York developer's candidacy would succeed.

S&P 500 futures slumped as much as 5%, hitting a trading limit, before slightly paring losses, as Mr. Trump pulled ahead of Democratic nominee Hillary Clinton. CME Group doesn't allow its stock-market futures contracts to drop more than 5% overnight.

Japan's Nikkei stock index was down 5.3%, crude oil slumped more than 3% and the dollar fell 3.6% against the Japanese yen, while the prices of U.S. Treasurys and other rich-country government bonds rose, sending yields down. The dollar was up as much as 13% against the Mexican peso. Gold, a haven, was up almost 5% at one point.

Tuesday evening's reversal set in stark relief the markets' fear of a possible Trump presidency. During his campaign he has advocated for sharply controlling immigration and raising tariffs on trade, decisions that many economists contend could pressure growth at trading partners such as Mexico and potentially reduce already sluggish global growth.

Many investors also are uneasy about the lack of specifics underlying some Trump proposals and the economic and geopolitical uncertainty that they contend would arise as a result of his presidency.

Please see REACTS page A12

◆ Global bond yields shrink, expectations for interest-rate hike take hit ... C1

Forecasts Missed Breadth of Support

Republican Donald Trump won the 2016 election, becoming the nation's 45th president in one of the biggest upsets in U.S. political history.

Mr. Trump reached his stunning triumph over Democrat Hillary Clinton in the early morning hours Wednesday, after a cliffhanger night of vote counting that capped the bitter, monthslong presidential contest.

By Janet Hook, Colleen McCain Nelson and Beth Reinhard

Defying polls and political pundits, Mr. Trump ran up extraordinary vote totals in rural areas and working-class counties across the U.S., where his message of economic populism resonated. It was an appeal the first-time candidate used in the

Please see VOTE page A8

'Deplorables' Rise Up To Reshape America

By Gerald F. Seib

ANALYSIS

The deplorables rose up and shook the world.

"Deplorables" was, of course, the disparaging term Hillary Clinton at one point applied to some supporters of Donald Trump.

Many of his loyal followers proudly embraced the insult and used it as a motivating tool.

Wearing such establishment disdain as a badge of honor, the Trump army cut a deep swath through the American electoral system Tuesday, propelling the Republican nominee to a stunning victory.

In the process of exceeding virtually all expectations, Mr. Trump has remade the Republican party in his own image. He rewrote some of the GOP's

Please see RISE page A8

ELECTION 2016

Exit polls find voters show a deep hunger for change

A2

Businesses grapple with new populist landscape

A3

Republicans maintain majority in Senate

A4

See WSJ.com for the latest vote results and election coverage

A Campaign Unlike Any in Modern Times

By Michael C. Bender And Peter Nicholas

In his last flurry of campaign stops, the Republican nominee Donald Trump began to exercise more discipline. Finally, it seemed, months after the campaign kicked off, after his impromptu Twitter tirades, the accusations of sexual harassment, the off-the-cuff comments and the uneven debate performances, his advisers had gotten control of their unwieldy boss.

Not so, according to Mr. Trump.

"I do what I want to do," he said in an interview over the final weekend. Mr. Trump predicted an "unprecedented victory," after which he would celebrate for "about an hour. Then I'll get up Wednesday morning, and start working so hard immediately."

Few believed him. Polls showed Democrat Hillary Clinton with a slender but stabilizing lede. Her party talked of taking back the Senate. The election, an unedifying, raucous and unpredictable contest, the strangest in the modern era, defied all the predictions.

At the New York Hilton Midtown, where the Republican nominee watched election returns, cheers went up from throngs of men in suits and women in red cocktail dresses every time a state was called for Mr. Trump.

An eager volunteer handed out red "Make America Great Again" hats while supporters posed for pictures against giant campaign-branded backdrops.

"It's starting to feel electric like it does at Trump rallies," said New York City lawyer Mark Smith. As it got closer to midnight, a cake in the shape of Mr. Trump's head was wearing a campaign hat.

Mr. Trump watched from his campaign headquarters "war room." As he has been from day

Please see RACE page A10

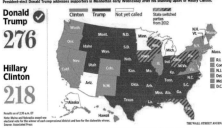

2016 ELECTION: WORLD REACTS TO DONALD TRUMP'S STUNNING UPSET VICTORY

f SHARE 406 | EMAIL | g+ SHARE | TWEET

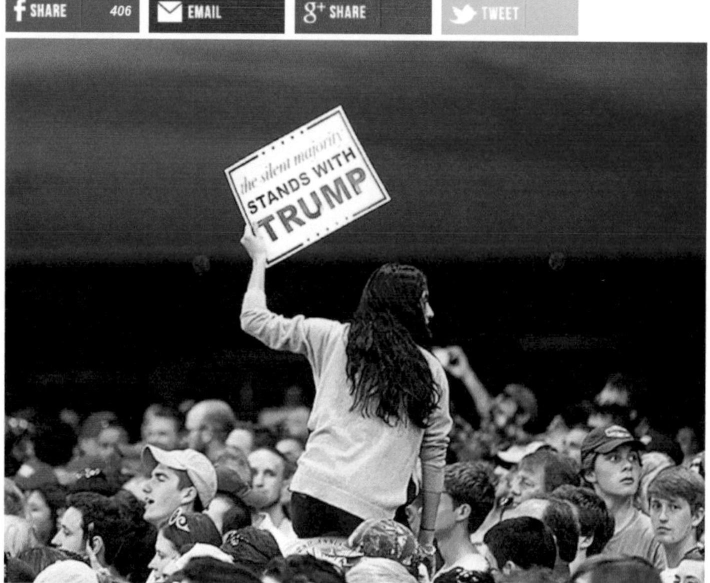

by BREITBART NEWS | 9 Nov 2016 | 23,461

Welcome to Breitbart News's live coverage of the aftermath of Election Day, 2016. Check this page for updates on president-elect Donald Trump, celebrating a shocking upset win, and the reactions from Democratic candidate Hillary Clinton, the bipartisan political establishment, and legacy media outlets.

All times Eastern.

7:13 PM — Mexican newspaper not taking this too well. Via elgrafico.mx:

BREITBART LIVE UPDATES

NECESSARY: 'CAN'T STUMP THE TRUMP' TAKES A VICTORY LAP

NIXON TO TRUMP: 'WHENEVER YOU DECIDE TO RUN FOR OFFICE YOU WILL BE A WINNER!'

AMERICA ON EDGE: MOVEON ORGANIZES ANTI-TRUMP PROTESTS AROUND COUNTRY

4.4

Breitbart web page story from November 9, 2016, captured from http://www.breitbart.com/.

the details of which I won't delve into here. Neither offers a direct means of downloading news coverage over an extended period of time.

These three sources are of interest because of their historical and material differences. The latter has to do with the "materialities of information representation" touched on in chapter 1, which Paul Dourish explains as the particular forms that information takes, and their implications for interpretation and action.[5] None of the outlets used here are what we typically think of as "local news"; they don't present a regional perspective. They nevertheless report the news of the day in varied ways: CNN is a twenty-four-hour television news channel (figure 4.2); the *WSJ* is a print and online newspaper with, as it turns out, deep historical ties to computational linguistics (figure 4.3); and BNN is a born-digital platform composed in the language of the web (figure 4.4). These differences shape the way that the news sources appear materially on the pages of this book as data. For each source must travel a distinct computational pipeline to finally arrive on the word list presented here. These pipelines are formed around the contours of each source, but also in relation to the NewsSpeak algorithm: the rules of the word list written in computer code. It is my hope that in seeing the news through the lens of NewsSpeak, readers might come to understand the fourth principle of the book: *data and algorithms are inextricably entangled.*

HOW DOES THE NEWS BECOME DATA?

Unlike purpose-built data, made in the care of scientific or historical collections such as those mentioned in previous chapters, what we might call "news data" is a product of "datafication": a process by which any incident or media might be quantified and translated into a form recognizable by computers.[6] Datafication is one of the primary mechanisms in the assembly of big data, and it has been used to digitally capture information about many of our social behaviors: our everyday movements are monitored by the handheld devices that we carry with us; our spending patterns are registered at the checkout line; and our reading habits are tracked online. In considering the results of datafication, as in other examples explored throughout this book, we must come to terms with their locality. How can the news—gathered from a variety of discordant and locally defined broadcast, print, and born-digital sources—become quantifiable and comparable as data?[7]

Understanding the locality of the news requires venturing into the places where the news is composed and viewed. But comprehending the locality of *news data* means something more: grappling with the algorithms that make the news accessible and manipulable, indeed functional, as data. Communication theorist Tarleton Gillespie writes of algorithms as "encoded procedures for transforming input data into a desired output based on specified calculations."[8] In practical terms, however, making the news into data means rendering it accessible to algorithms, like the ones that I used to produce NewsSpeak. My algorithm is intentionally simple so that readers can understand its mechanics and their relationship to different data sources.

In the *Language of New Media*, Lev Manovich explains algorithms and data as symbiotic. One cannot exist without the other. "Together, data structures and algorithms are two halves of the ontology of the world according to a computer."[9] And yet up until this chapter, I have discussed data independently of their other half. Now I expand my exploration to include the implications of locality for algorithms. In the context of new concerns about their potential biases, algorithms are rapidly becoming the subject of widespread public interest.[10] But we cannot simply think of data and algorithms as independently originating elements of computing. Specific data sets and algorithms, designed to work together, cannot be easily separated and appropriated for other ends.

Following on the 2016 election, understanding the locality of news data in the US has never been so important. First, election coverage now unfolds online at a more rapid pace than ever before, demanding our continual, twenty-four-hour attention. Second, since their failure to call the election, news sources have been under scrutiny from all sides.[11] Federal investigators claim that a foreign power, Russia, might have introduced "fake news" that influenced the vote.[12] Trump, meanwhile, routinely accuses the establishment media of being fake.[13] As a result, many Americans exclaim that they no longer trust any news media. Third, media companies such as Facebook, Twitter, and Google have been asked by the US Congress to reflect on the role that their algorithms play in curating our everyday engagements with the news.[14] That work is long overdue, for the relationship between news, data, algorithms, and reality, as currently configured by such companies, is getting more confused by the day.

My efforts to parse the election coverage as data required access to not only a wide range of news data but also the algorithmic tools for natural language processing (NLP). This is an increasingly prevalent set of algorithms created to recognize and manipulate everyday discourse.[15] NLP tools support all kinds of computer manipulations of language, from counting word frequencies to comparing writing styles. Ultimately, researchers in NLP endeavor to help computers understand complete human utterances, at least well enough to provide useful responses. The successes of NLP research can already be seen in software for language translation and digital personal assistants like Apple's Siri. In order to create user experiences that feel more intuitive, indeed more natural, developers use recent advances in machine learning: algorithmic techniques devised to train computers to recognize expected patterns of speech, using large example data sets rather than explicit rules.[16]

But let us dig beneath the placating surface of such applications in order to understand the material and historical assumptions of NLP. I built NewsSpeak as an exploratory exercise. Indeed, I found that its constitutive elements—data, algorithms, and assumptions about news media and even its connection to reality—had to be continually reworked in the process of creating it.[17] The limits of NewsSpeak, a function of the incongruent textual structures that I discovered in the news sources, then led me to an inquiry into the historical relationship between NLP algorithms and news data.

Too often, algorithms and data are discussed as related but nevertheless discrete elements of computing. I argue that being able to identify and analyze the local conditions that shape news data is crucial for understanding related algorithms as well as the limitations inherent in applying them as generic tools. Looking at the news as data reinforces a recurring theme of this book: we cannot continue to think about data sets as autonomous units to be taken up indiscriminately in practices of computing that have broad implications for society.

CREATING NEWSSPEAK

In spring 2016, I began to compose the NewsSpeak algorithm as a form of inquiry into the Natural Language Toolkit (NLTK), a widely used library for NLP, and one that is representative of a larger set of toolkits.[18] As part of that endeavor, I also had to work with Python, a computer language necessary to make use of NLTK. Along the way, I discovered that the news and NLTK articulate one another in distinct ways dependent on local formations of each. Moreover, I learned that the news has played a crucial role in the history of NLTK, which I will explain in due time. Let us start with the details of the NewsSpeak algorithm. Below is an outline of what the algorithm does, written in pseudocode: a kind of natural language summary of a computer program:[19]

```
GIVEN A CSV OF ROWS IN THE FOLLOWING FORMAT: 'date,' 'outlet,' 'sentence'

CREATE A UNIQUE_WORDS DICTIONARY
CREATE A PREVIOUS_WORDS LIST

FOR EACH ROW IN THE CSV FILE:

   SET DATE TO THE FIRST ELEMENT IN THE ROW
   SET OUTLET TO THE SECOND ELEMENT IN THE ROW
   SET SENTENCE TO THE THIRD ELEMENT IN THE ROW

   SPLIT SENTENCE INTO INDIVIDUAL WORDS IN A WORD_LIST

   FOR EACH WORD IN THE WORD_LIST:

      IF THE WORD IS 'ELECTION' AND NOT THE FIRST WORD IN THE WORD_LIST:

         SET PREVIOUS_WORD TO THE WORD BEFORE 'ELECTION' IN THE WORD_LIST

         IF PREVIOUS_WORD IS NOT ALREADY IN THE LIST OF PREVIOUS_WORDS:
            MAP PREVIOUS_WORD TO AN EMPTY INSTANCE_LIST IN UNIQUE_WORDS
            ADD PREVIOUS_WORD TO THE LIST OF PREVIOUS_WORDS
```

```
        CREATE A DATE_AND_OUTLET DICTIONARY
        MAP 'DATE' TO THE DATE
        MAP 'OUTLET' TO THE OUTLET

        ADD DATE_AND_OUTLET TO THE INSTANCE_LIST MAPPED TO PREVIOUS_WORD

FOR EACH WORD AND INSTANCE_LIST MAPPING IN UNIQUE_WORDS:
   SORT THE DATE_AND_OUTLET'S IN THE INSTANCE_LIST BY DATE IN DESCENDING ORDER
   SET FIRST_USAGE TO THE FIRST DATE_AND_OUTLET IN THE SORTED INSTANCE_LIST

   SET UNIQUE_WORD TO THE WORD
   SET USAGE_DATE TO THE VALUE OF 'DATE' IN FIRST_USAGE
   SET OUTLET TO THE VALUE OF 'OUTLET' IN FIRST_USAGE
   SET USAGE_COUNT TO THE LENGTH OF THE INSTANCE_LIST
   SET DAYS_PRIOR TO 11/8/16 - USAGE_DATE

   WRITE WORD, USAGE_DATE, OUTLET, USAGE_COUNT, DAYS_PRIOR TO A CSV FILE
```

The purpose of the NewsSpeak algorithm is threefold: to introduce readers to the basic mechanics of NLTK using data from a three-part corpus (CNN, *WSJ*, and BNN) that they are likely to be familiar with; to reveal how news data can be understood locally in relationship to the conditions of their creation and the site of their use (for instance, to create an output that fits on these printed pages); and to show how a data set, algorithm, and implicit representation of reality (that is, what the data can be used to argue) must be constructed in tandem.

My work on NewsSpeak began with sample data exported from the NewsScape archive. As an initial source, I choose to work with CNN coverage because of its accessibility via NewsScape along with its breadth and prominence within the 2016 election cycle. There are several means of accessing CNN data through NewsScape. I made use of the basic search interface that allows keyword queries filtered by organization, date, and other variables. Such search queries in NewsScape are applied to existing closed-captioned text for the news, created in the course of each broadcast.

From the outset of this work, I found the news snippets based on closed captioning to be a compelling way of excerpting the news, though with significant limitations: the original video and audio is missing, and along with them the tone and imagery that add layers of context, particularly useful in understanding the never-ending drama that characterizes twenty-four-hour television news.[20]

Yet even putting aside the missing video and audio, closed-captioned text is shaped by its own local conditions related to both the broadcast and transcription processes. Transcriptions today are created through a human-machine process, which relies on voice recognition software, but not in the way that you might think. Straight audio from a broadcast is too complex for current voice-to-text software to translate directly. Instead, human transcribers act as intermediaries, repeating (almost) everything said on-screen into a computer-connected microphone. They must enunciate in just the right way, and use command words to add punctuation as well as macros: verbal shortcuts, programmed for anticipated words and phrases that are difficult or time consuming to verbally encode.[21]

Frequently there are errors. Words are left out or mispronounced, and in turn misspelled or substituted by similar-sounding words that the system has been trained to expect. Sometimes the macros are the source of errors.[22] This work can be enhanced if transcriptionists have access to the scripts created ahead of time for newscasters. But the most urgent "breaking news" reported on the fly or from the field does not follow a script. As a result of all these contingencies, the corpus of closed-captioned text that NewsScape uses to search and share broadcasts is inherently uneven.

Early on in my examination of NLP, while using NLTK functions to explore coverage from NewsScape, I became aware of another subtle though important material way in which the natural language algorithms process news. When the techniques of NLP are applied to the news, time is flattened out. I spoke about the strange temporality of the news with Sergio Goldenberg, the director of technical product management for TV and mobile applications at CNN. He explains that we are meant to encounter the news only in the present. Every news story is broadcast, printed, or posted with the understanding that it is self-contained and ephemeral. Newsmakers cannot assume that audiences will have read or heard yesterday's news or that they will return for a follow-up story tomorrow.[23]

There is an old saying, notes Goldenberg, that "today's news is tomorrow's fish and chip papers," which both acknowledges the temporality of newspapers and puts them in perspective. Likewise, on television and radio, news happens in the moment. Archives of the news, like NewsScape, change the temporality of the news. They rely on algorithms to make news data part of the retrievable record of what has been said by journalists and their subjects across time.

Thus what originates as a sequence of statements unfolding in real time is stored by NewsScape as one contiguous list of words. The past and present now share the same frame. Only by making the news into storable, searchable, retrievable data, through the techniques of NLP, is the temporality of the news redefined as something that accretes.[24] This early lesson—that the datafication of the news challenges our temporal expectations (it's no longer just a representation of what's new)—stayed with me throughout the NewsSpeak exercise. Building on this conception of the news as something additive, my work with NewsScape, which up until this point had been broadly exploratory, started to converge on a question: Can an algorithm reveal what's new in the news?

As I worked, my data set shifted and expanded to encompass different sources. I experimented with subsets of CNN, such as individual news programs as well as different query terms and time periods. In parallel, I had to create new algorithms to augment or supplement those offered by NLTK. This was a messy process with many false starts, wrong turns, and abandoned efforts. Suffice it to say that my algorithms had to stay in sync with the changing data that I was working with. For example, before settling on the search term *election*, my earliest implementations of NewsSpeak focused on the names of presidential candidates or issues related to the election. Furthermore, I made a deliberate choice to focus on a single term (or "token" in the language of NLP). I did this for the simplicity of the algorithm, leaving aside other related terms like *race* or *vote*. I also had to decide how much data to take as input for the algorithm. As I explored various parameters for the data set, I was forced to think about my "what's new?" question in material terms. For instance, the number of days included and the page space in this book were formative variables.

Meanwhile, I was discovering just how different various news sources were in terms of their structure. Rather than seeing the limits of NLTK as a challenge, I viewed them as an opportunity to think about how close relations between data and algorithms might be examined. As mentioned previously, I added two other outlets to my investigation, from outside the confines of NewsScape: the *WSJ* and BNN.

The *WSJ* is a daily newspaper, first published in 1889 to handle business and financial news.[25] I attained digital *WSJ* coverage of the election from ProQuest, a service that requires a paid subscription. I choose the *WSJ* partially because of its historical ties to NLTK, which I will revisit later in the chapter, but also as a prototypical example of print news.

BNN has only recently gained widespread attention, in large part because of its role in the election. It is a favored news source for Trump, often inspiring his widely followed comments on Twitter. Trump's onetime chief strategist, Steve Bannon, long served as the head of the online news organization. BNN is important for another reason, though. It is an illustration of a born-digital news source—one that is structured by tags and links, as opposed to newsprint pages or broadcast times.

Taken together, these three sources certainly do not tell us everything about how discourse in the news might be made into and managed as data. But they offer a comparative understanding of how the news, created in multiple formats (each with their own constraints), becomes data. Developing an algorithm to bring together these discordant sources required a number of nested decisions about what to include, how to focus, and what to leave out for clarity. After all, NewsSpeak is not meant to be innovative, but rather instructive. If it strikes some technical readers as a routine use of NLTK, that is intentional.

Before I consider the implications of this work, let me summarize. The current version of NewsSpeak is assembled using code written in Python. It takes as input news from three sources, beginning on September 16, 2015.[26] Each of these news sources is processed through the same algorithm, which searches for the string *election* in the selected coverage (although this is done in different ways for different sources). The algorithm identifies the string directly before *election*, filtering out a variety of conditions (i.e., if the string is a "."). These exceptions are sometimes known as "stop words." If the string is new—if it has not been seen before by the algorithm—it is added to a collection of strings that are kept in the order in which they first appeared. If the string is not new, it is added to an internal tally, recording the number of times that string has been used. The collection is presented on the pages of this book in graphical form, as a list of terms that modify *election*, in order of use leading up to the day of the vote on November 8, 2016. JavaScript was used to make the resulting text list visible in graphical form, adding color for each outlet and appending the number of times that the word was used in each.

Observations from NewsSpeak

The NewsSpeak algorithm shows the news as NLP sees it: as a list of words. The most apparent observation from the list is the dominance of CNN in relation to the *WSJ* and BNN. But this should not come as a surprise to anyone who knows these outlets and the varied frequencies at which they do their reporting. CNN broadcasts news continuously, while the *WSJ* and BNN are more periodic in their output. If I were performing a legitimate discourse analysis of the news, I would probably normalize these numbers—a process introduced in chapter 3, helpful for comparing different sources on an equal basis. Yet that is not my intention here. Rather, I want to reveal what routine practices, such as normalization, conceal.

Other reflections from a high-level, quantitative view are of limited value too. The most commonly used modifiers to the term *election* are not informative: *the* (mentioned 8,832 times on CNN, 62 on the *WSJ*, and 345 on BNN), *this* (6,280 CNN, 24 *WSJ*, and 85 BNN), and *general* (6,248 CNN, 9 *WSJ*, and 202 BNN). One can make some interesting observations, however, by reading a little more closely. For instance, CNN tends to do the most editorializing. It uses the highest number of overtly negative and positive modifiers to the term *election*. The positive terms *exciting* (12 CNN), *happy* (8 CNN), and more ambivalently *clear* (40 CNN) are all used by CNN exclusively. *Healing* (12 CNN) is perhaps the most noteworthy positive modifier in that it was used twelve times in the two days leading up to the election. Meanwhile, CNN also led in the use of negative terms. *Rigged* (522 CNN, 3 *WSJ*, and 2 BNN) was the most commonly used negative modifier, by an order of magnitude. It was also one of few terms used by all outlets. But CNN was alone in its use of many other negative terms, such as *stolen* (21 CNN), *nasty* (16 CNN), *crazy* (19 CNN), *divided* (18 CNN), *ugly* (11 CNN only), and *sad* (11 CNN), just to name the most commonly used illustrations. The *WSJ* didn't seem to use *election* much at all. When it did employ the term, its preceding words, such as *municipal* (3 *WSJ*), are descriptive not editorial. Meanwhile, BNN was the first to bring up many past dates, suggesting that its narratives about the election might have a historical arc.

Such observations are of limited value, however, because of the constraints of the algorithm, which reads the election coverage selectively. Typically, this kind of word frequency counting algorithm would be the first step in a lengthier program, designed to draw quantitative conclusions about these news outlets, based on the words that they use. For example, NewsSpeak could be linked to sentiment analysis, a technique for determining whether a corpus is generally positive or negative. Such an approach could expand our understanding of the differences among these three outlets or suggest how the general sentiment of the news changed over time as the election day approached.[27] Alternatively, we could assess the partisanship of the news outlets by comparing the words in this list to known instances of partisan discourse. But I will not follow either of those models, or unpack the equally contingent algorithms that they necessitate. For as I have already indicated, my purpose here is not to directly analyze the news.

NewsSpeak is meant to create an experience of news data on these pages that we can reflect on in material terms: colored and composed for the constraints of book printing, with special relevance for audiences that followed the election as it unfolded in real time. For people who lived through the election coverage, the list acts like an index, prompting them to revisit the election in a new form, and retrace its discourse over the months leading up to the final vote. This experience might be different for each reader. After all, NewsSpeak is not meant to reveal a single pattern or conclusion. Instead, it offers a reframing of the news as data.[28] It allows us to experience the election in a novel way, enabled by a systematic computational filtering of recorded election coverage. When understood as such, the NewsSpeak algorithm reveals itself as an aesthetic encounter with the news, more akin to a poem or performance than a political analysis. It is meant to raise questions about what words do in the news, what they don't or can't do on their own, and how they might be reorganized for alternative reading experiences.[29]

Data Artifacts in NewsSpeak

What can this list of words tell us about the ways in which data and algorithms are entangled? Below I address a handful of localized incidents—referred to as *data artifacts* in previous chapters—that are not immediately apparent from the list, yet should inform our reading of it. These data artifacts are symptomatic of broader challenges in NLP that ongoing research in computer science is trying to address or circumvent.

First, the list obscures the high number of transcription errors in live news reports. Take row 25, for example. The term *welcome* does not seem out of place. But consider that it comes from this sentence:

> I learned that we have a lot of talent in the republican party and I think we're going to do very welcome election day. (CNN, September 17, 2016)

Reading the full sentence, it is easy to see that it should say "very well, come election day." This is likely a mistake caused by rapid, on-the-fly transcription. I spoke with a professional transcriptionist, who explained that broadcast news presents special problems, illuminated in artifacts like this one.[30] As mentioned earlier, broadcasts must be transcribed in real time, commonly without a script, which is likely to lead to such errors. Often these errors are linked to the use of new words or phrases, presenting a special issue for my algorithm as it is attempting to portray the first uses of words to describe the election. New words or phrases may be mistyped the first time simply because transcriptionists are not prepared to hear them. Moreover, many standard phrases are programmed in, which makes them easier to use. Of course, these are the first uses in NewsSpeak, during a certain time period and related to the election, so they are not new words in the larger sense.

A different set of problems accompanies work with text created to be read online. Take, say, the string below from BNN, identified by NLTK as a full sentence. In this excerpt, a period (preceding and not present in the string itself) is interpreted by NLTK as an indicator of the end of the previous sentence. In actuality, it is part of a uniform resource locator, an example of nonnatural computer language or, in this case, HTML.

> com election-tracking site. (BNN, September 20, 2016)

The list of terms also doesn't help us recognize whether a term is being used literally or as a metaphor. Consider the use of the term *school* below. In this instance, the commentators are referring to the election at hand, but only obliquely. For they are comparing it to a high school election.

> So getting small, getting—you know, making it look like it's just a high-school election and name calling and all of those things, that can actually take you out of the campaign tonight. Stay big. I think that's one of the things we want to see. (CNN, September 15, 2016)

The hyphen used above is not picked up by NLTK (a common error), so we only see *school* on the list. Another problem involves identifying complex nested quotes, and who ultimately is being quoted. In an entry to the list from a CNN broadcast, *weird* refers to a term that Trump used, not commentary from a journalist:

> Stuff with jeb it gets covered because trump is the candidate that the media is more obsessed with than anyone else. >> i love that he says this is a weird election as if he's removed has nothing to do with the fact that this has been a weird election. >> yeah thanks to trump it's weird right? >> it is a very weird. (CNN, November 10, 2015)

As in the case above, journalists are frequently analyzing the speech of others, complicating efforts to assign attribution. Finally, there is the problem of identifying scope. Are journalists talking about the 2016 US presidential election or another one? The terms *endless* and *cliff-hanger* are about the 2000 election, when George W. Bush defeated Al Gore, and *supercompetitive* refers to the 2008 elections. Also, I enjoyed discovering the following humorous example:

> SIGN UP FOR OUR NEWSLETTERA man disguised as the Star Wars character Chewbacca was arrested in Ukraine for breaking election laws after he drove â€œDarth Vaderâ€ the Internet Party candidate for mayor of Odessa to vote in local elections. (BNN, October 26, 2015)

Because of these artifacts and others, many forays into the analysis of news data collections use methods other than NLP. The Internet Archive, a nonprofit library of internet sites, has used face recognition in order to track reoccurring news coverage of individuals.[31] Media Cloud, an "online platform for media analysis," has made effective use of network visualization of the news using the identifiers of news sources alone.[32] Another project, titled "Page X," adapts a traditional form of news analysis—how much "real estate" on a page does a story get from different outlets—for born-digital news.[33]

Using NLP to effectively analyze the news at a large scale is not that practical today. In fact, it is an unresolved research problem, but an active one. For instance, the NLP community is currently using open competitions to synchronize efforts to improve automatic news comprehension. One recent example targets the identification of fake news—a problem identified by the 2016 election.

Were US flags "banned from display at the 2016 Democratic National Convention"?[34] Did Trump once exclaim that Republicans are the "Dumbest Group of Voters"? Was a pedophile ring "operating out of a Clinton-linked pizzeria called Comet Ping Pong"? None of these headlines are true. But such illustrations of fake news, or simply propaganda, might have influenced the 2016 election by reaching millions of users on Facebook and other social media sites.[35]

The NLP Fake News Challenge proposes that we use algorithms to identify and dethorn fake news stories before they can do more harm. Although this challenge represents the highest ambitions of the NLP community, it is articulated as a problem of aiding human analysts rather than replacing them:

> The goal of the Fake News Challenge is to explore how artificial intelligence technologies, particularly machine learning and natural language processing, might be leveraged to combat the fake news problem. We believe that these AI technologies hold promise for significantly automating parts of the procedure human fact checkers use today to determine if a story is real or a hoax.[36]

Indeed, today's algorithms are not actually autonomous; they must work in tandem with human users and preexisting data sources. Developing NewsSpeak helped me come to terms with some of the resistances involved in bringing together unrelated data and algorithms. I had to wrangle the data and algorithm in tandem. As I modified the data set—by changing the sources, scope, or search terms—the associated algorithms had to be rewritten or repaired accordingly. Data, by virtue of their representation in linguistic form, cannot simply be plugged into existing NLP toolkits.

This kind of necessary tinkering, with both data and the algorithms, is a mundane yet necessary part of data work. By many accounts, practitioners spend the majority of their time "cleaning" the data or "debugging" algorithms. But such language, which suggests that any problems in either data or algorithms come from the outside—from errant data dirt or computer bugs, which have infiltrated an otherwise-pristine system—obscures the complex local conditions in which both data and, by extension, algorithms are made.

Rather, local sources of the news and NLP algorithms vary in their assumptions about what constitutes "natural language." Those assumptions are best understood by examining the history of engagements between algorithms for NLP and various data sources. In their early development, the algorithms that make up the the NLTK library and in fact all NLP toolkits were trained on particular formulations of natural language.

HISTORICALLY NATURAL LANGUAGE

The history of NLP offers a striking example of how algorithms develop in conversation with research goals as well as changing ideas about what counts as "good" data. Algorithms and data shape one another, not only in projects like NewsSpeak, but over the course of long-term efforts in research and development. Up until 2001, the evolution of NLP could be explained in four distinct phases, according to Karen Spärck Jones, one of the most prominent researchers in the area and a former president of the Association for Computational Linguistics.[37] Each phase was grounded in the local conditions of its time.

The first phase ran from the 1940s to the 1960s and focused on machine translation. Although work on translation was international from the start, funding in the United States concentrated on Russian-English translation and was contingent on an encompassing Cold War political setting, which provided the motivation and funding from the military. Due to their practical defense-related needs, translation algorithms were simple at first, revolving around on word- or sentence-level discourse.

Work during the second phase shifted to pick up momentum from early robotics work in the late 1960s and 1970s. During this time, NLP was framed around the problem of manipulating knowledge representations. In a project called SHRDLU, a simulated robot designed to manipulate blocks in a computer model could carry out basic commands, such as "pick up the red pyramid." The project, led by artificial intelligence pioneer Terry Winograd, demonstrated that a computer could understand natural language, albeit in a highly restricted and quantifiable domain: inside the curated virtual space of a computer. Any attempt to change the setting or scale of the activity resulted in dramatically reduced effectiveness.[38]

The third phase, from the late 1970s to late 1980s, was a time of significant progress for grammar theory. A focus on semantics led researchers to investigate how meaning inevitably referred to the context of discourse and an encompassing "world model." Most notably, this era produced some of the first practical and more widely accessible toolkits, such as the Alvey Natural Language Processing Tools.[39]

Only in the fourth phase, beginning in the early 1990s, were the tools developed in the previous eras turned toward the analysis of large corpora of text. Finally, the news became a prominent source for the development of NLP algorithms. Early on, processing the news centered on *information extraction*, which can be defined as identifying predefined classes of terms within existing texts, ranging from syntactic entities, such as parts of speech, to semantic entities, such as subjects and objects of action.[40] As in previous eras, the most effective systems were locally constrained: they could only function in narrow, predefined settings. A veteran researcher in computational linguistics, Charles Fillmore, explains an early system designed to read the daily news:

> I witnessed work on information retrieval in the form of a system that automatically collected information from newspaper accounts of traffic accidents. My impression was that the system was given texts that were known to be about traffic accidents and it was already provided with a checklist of information to look for, based ultimately on the style sheets used by reporters working on traffic accident assignments, or, really ultimately, on the reporting traditions of the local police departments.[41]

These limitations were widely acknowledged within the NLP community. Jones notes that the most effective systems at the time of her writing were those with the simplest tasks and most tightly constrained domains. Jones's story ends in 2001, the same year that NLTK was created for a computational linguistics course in the University of Pennsylvania's Department of Computer and Information Science. Today NLTK is freely available online. It has become one of the most popular computational toolkits for working with language in a range of applications including "predictive text and handwriting recognition," commonly found in messaging applications, and "web search engines that give access to information locked up in unstructured text."[42]

Like other programming toolkits, NLTK is not a single algorithm but instead a library: a collection of prewritten code that can be invoked by a programmer in order to create their own software. Libraries contain multiple algorithms. NLTK is based on statistical models developed in order to address the limitations of brute force algorithms used in earlier eras. This probabilistic approach, informed by grammatical theory, has given rise to systems that can perform work that even native speakers sometimes find tricky, like identifying parts of speech.[43] NLTK, however, is limited in its handling of language. It is a good illustration of how all algorithms are trained or at the very least tested on local, sample data sets, which fundamentally shape their normative assumptions: what the algorithms should and can be used for.

As such, libraries for NLP such as NLTK are not neutral or general. They do not offer a complete and unbiased comprehension of language use. Computer scientist Joseph Weizenbaum once wrote, and I believe it still holds, that there is no such thing as a general language processor, not even a human one. He invented an early chat bot, named ELIZA, programmed to parody the role of a psychotherapist. Weizenbaum was deeply disturbed by what he saw as unearned reactions to the algorithms that drive ELIZA: "A widespread, and to me surprising, reaction to the ELIZA program was the spread of a belief that it demonstrated a general solution to the problem of computer understanding of natural language. ... [E]ven people are not embodiments of any such general solution."[44]

But NLP tools are also not "inhuman" or "alien," as they are sometimes characterized.[45] Rather, such algorithms are themselves local, for they are created and applied within particular historical and material conditions. For example, data based on the *WSJ* provides a basis for NLTK.[46] As such, the toolkit relies on the insufficiently acknowledged labor of journalists and editors whose English language knowledge has now been subsumed by the algorithms. Furthermore, countless person-hours were contributed by highly trained linguists who annotated the original corpus, converting it into a form that existing machines can process. We might say that NLP is actually a type of human reading—one that is mediated by an increasingly sophisticated system, which masks the work of its many individual contributors.

Let me explain how this particular set of algorithms came to be entangled with an oddly specific data setting, defined by the *WSJ* and its human creators. NLTK relies on a kind of machine learning, trained using the Penn Treebank, a corpus of over 4.5 million words in American English first created in 1989, and painstakingly annotated for both syntactic and semantic structures by expert linguists.[47] A treebank attempts to represent the hierarchical treelike structure of language. It is a "bank of linguistic trees."[48] In the Penn Treebank's most recent manifestation, the primary component of this corpus is a million words from editions of the *WSJ* published in the late 1980s.[49] Tagging this corpus consists of identifying distinct grammatical behaviors within it and ideally coding each word with a single behavior (that is, *VB* for infinitive or imperative verbs, and *NN* for singular common nouns). Authors of the NLTK book, a widely used resource for learning NLP, explain that "the training process involves inspecting the tag of each word and storing the most likely tag for any word in a dictionary."[50] NLTK has a variety of automatic tagging algorithms, each of which accounts for the linguistic context of a word in different ways. Only by determining the proper context, an ongoing research project in computational linguistics, can algorithms determine how a word should be categorized.

As a means of further understanding NLTK, it might be helpful to revisit some of the problems introduced in the example algorithm, NewsSpeak: mistakes in transcription, misinterpretation of punctuation, and a general lack of context. But NLTK researchers are perhaps most challenged by ambiguities or double meanings, which are an inherent part of language. They often trip up human readers, so how can we expect computers to do any better? Jokes frequently play on such ambiguities. The authors of the Penn Treebank demonstrate this using a humorous, if somewhat dated, quip from the actress Katherine Hepburn, presumably about Cary Grant, with whom she made several films: "Grant can be outspoken—but not by anyone I know."[51] Here the term *outspoken* will likely be interpreted by an NLP "part of speech" tagger as an adjective, even though, in the context of the joke, it is meant to be recognized as the past participle of the verb *outspeak*. Hopefully some readers will get this and be amused. Such jokes, however, can easily miss their mark with an unprepared audience. Many younger readers might wonder, who is Cary Grant?

The Penn Treebank contains thirty-six parts of speech tags, and twelve tags for punctuation and other symbols—a dramatic reduction and streamlining of previous models. The annotation was originally performed through a combination of machine tagging and human correction, which was found to be considerably faster than human annotation alone. Nevertheless, the cost of annotation, which requires a high level of expertise, is prohibitive, and one of the major reasons that the Penn Treebank has not be updated often or replicated.[52]

When they wrote "rapid progress can be made in both text understanding and spoken language understanding by investigating those phenomena that occur most centrally in naturally occurring unconstrained materials," NLTK's creators did not necessarily intend *naturally occurring* to mean a set of completely generalizable examples of American English.[53] Researchers are much too nuanced for that.

Early on the Treebank processed a number of other specific corpora, including WBUR transcripts (an NPR-affiliate news radio station based in Boston), IBM computer manuals, and Library of America texts, the last of which consist of small passages from US-based authors like Mark Twain, Herman Melville, W. E. B. DuBois, and Ralph Waldo Emerson. All these choices reveal much about the expectations among NLTK's designers as to what are model forms of language use: examples considered canonical or commonplace (all by men, most of them white). It was only later that they turned to the *WSJ* as an exemplar of everyday language use.[54]

Readers who are not part of the computational linguistics community might need to be attuned to some of the compromises implicit in the efforts to apply computers to the understanding of natural language. I do not suggest that NLP researchers are unaware of these concessions. Only that they shape what the rest of us must come to understand as a localized model of natural language, informed by circumstances that should not be overlooked. First, NLP accepts more or less that language can be disassembled into a list of words to be analyzed and manipulated as discrete, self-contained units of expression.[55] As such, researchers must accept a view of language as principally informational as opposed to poetic or even perhaps intentionally opaque, as in the case of lingo, jargon, and slang. Second, researchers have confined themselves to a certain type of grammatically precise language at a level only possible in print or scripted speech. NLP works on the supposition that language is textbook correct, and has been professionally edited for spelling and grammar. Third, language is expected to create meaning independently of other media, such as images, which are routinely part of the news (even print news like the *WSJ*). Finally, there is a regrettable assumption in the continued use of historical NLP resources like the Penn Treebank that the rules of language use do not change. Yet tags created in 1989 to describe articles from the *WSJ* cannot be effective indefinitely. Linguists know that even punctuation has not always been part of natural language use and continues to evolve, as the emergence of emoji use reveals.[56]

Many of these assumptions, necessary for NLP to produce results, do not hold in the majority of language use. Try running with these assumptions on Twitter, where rapidly changing forms of language use, such as the introduction of hashtags or the intentional appropriation of humorous errors, make NLP use difficult.[57] Nevertheless, they have been adopted, often consciously, as a part of the ongoing struggle to get computers to recognize our communications.

Current approaches to NLP are not wrong, of course. They are simply based on limited models of language use as well as situated human labor, much of which unfortunately goes unacknowledged. It is important to concede that NLP algorithms are shaped by the locality of data, for they hold significant implications for the future trajectory of public discourses in the news. Indeed, some news makers are already bending their products into forms that are more easily recognizable by current algorithms.

DATA, ALGORITHMS, AND THE REALITIES THEY SUPPORT

Only recently have algorithms been appreciated as subjects of broad relevance to everyday life. They introduce new and opaque procedures to important domains of public understanding and decision making, such as the news, finance, criminal justice, and even love.[58] Algorithms have become the matchmakers of our time: they illuminate connections across data from diverse sources. But algorithms are not just technical procedures. Social studies of algorithms have revealed them to be complex sociotechnical artifacts: fragile, multiple, and situated in ad hoc practices of computational work.[59] Algorithms are local, not in small part because they rely on data for their development and testing. I would take that argument one step further: collections of data and algorithms should not be considered as entirely independent components of computation. Indeed, they are entangled with each one another, materially and historically.

Let us return to a question from the beginning of this chapter: How does the news become data? We can now see that processes of datafication are enmeshed in complex local conditions. Searching for a way to characterize how data are made during the all-consuming rush to digitize media—images, videos, music, text, and more—starting in the 1990s, Manovich posited that a new "cultural algorithm" had emerged:[60]

reality -> media -> data -> database

Reality, explains Manovich, is represented through media, which are then digitized as data, and ultimately structured and stored in databases for future use. His prescient concern at the time was that data might eventually overshadow their original sources. For viral images, video, and texts had already spawned innumerable copies online. Invoking Jorge Luis Borges, a favored storyteller of media theorists, Manovich observes that data are a new means by which the map can exceed the territory.[61] Data, he argues, have become a dominant means of sense making, eclipsing direct engagement with reality.

This concern, however, presupposes a stable and singular reality that can be represented by media, and ultimately shoehorned into a database. We might call this a "realist" explanation of how data come to be in the world. When applied to our thinking about the news, the realist explanation suggests that the world is authentically captured by news media first, then later made to conform to the constraints of contemporary data and database structures, such as those that govern how news is shared on contemporary social media platforms. As for the role of algorithms in this explanation, that is left more abstract. Manovich presents algorithms as processes that govern the relationship between reality, media, data, and the database, but that nonetheless exist independently from them.

This may appear to be the case if we look narrowly at the way in which the NewsScape archive ingests the news into its databases. Yet I believe Manovich's realist characterization of data needs some rethinking. For example, it's not difficult to imagine how the arrows of Manovich's algorithm might go in the other direction:

database -> data -> media -> reality

Depending on whether the database is relational, hierarchical, or networked, its structure will shape the kind of data that can be created.[62] Data in turn can be used to create media in the form of visualizations, graphs, charts, and other forms of analysis around which conceptions of reality are ultimately built. As Jean Baudrillard asserts, the map can precede the territory.[63] Conceptions of reality must conform to predefined categories of data in order to be legitimized. Consider how NewsScape prioritizes closed-captioning text from broadcasts; it is easier to translate those elements of reporting into a searchable form. Meanwhile, the video and audio are given less attention. As another illustration, the reader may recall from the last chapter how a library must catalog its new arrivals into a predetermined system of organization, such as those established by the Library of Congress, Dewey decimal classification, or Dublin Core.

This reverse ordering of Manovich's cultural algorithm might be called a "constructivist" explanation.[64] In contrast with Manovich's realist version, it suggests that conceptions of reality are, as the name implies, constructed from explanatory media and based on data that are dependent on an underlying database infrastructure. In this formulation, algorithms are still independent, directing the progress from database to reality from some external position. Yet this explanation does not hold water either. From what is data derived if not reality? The constructivist explanation has no clear ground.

Whether it proceeds forward or backward, by realist or constructivist steps, Manovich's cultural algorithm is meant to explain how data, algorithms, and reality are connected. By calling this algorithm "cultural," Manovich gestures to the fact that a range of processes (not all technical) are at work, helping to create and manipulate data. Rather, I would like to suggest that unpacking the algorithm—understanding it in specific rather than general terms, as I have sought to do here—can help us see how various elements of computation coevolve in a parallel instead of sequentially. The relationships between data and algorithms are not generic. They are bound up in specific conditions, such as the choice made for the Penn Treebank to use the *WSJ* to shape contemporary NLP toolkits.

There is another kind of explanation, commonly employed in social studies of science and technology, exemplified by "actor-network theory" and "ontological politics," that can help us understand how reality gets done locally.[65] Reality is not universally fixed, these explanations assert. Rather, to adapt an assertion made by Bruno Latour, reality simply consists of the set of statements too costly to call into question.[66] Such "network" perspectives would hold that reality is not at the beginning of the algorithm (the input), nor its end (the output). In fact, statements about reality are another element at play in the continual search for an effective arrangement of media, data, databases, and algorithms. A network perspective on datafication assumes that all these elements coevolve:

reality + media + data + databases + algorithms

Significantly, network explanations imply that there are multiple realities, each resonant with an organization of algorithms, data, databases, and media. If we accept this view, we can move beyond the limits of the realist and constructivist explanations. The elements of a network are not discrete, linear steps to be connected by an algorithm.[67] Rather, these elements and the relationships between them are perpetually under construction. What's more, algorithms and even reality are themselves part of the arrangement.

Bringing a network perspective to news data can help us understand datafication with tools that are useful for thinking about the past, present, and even future of the news. This perspective helps us describe a condition that the material and historical inquiries earlier in the chapter support: news media, data, databases, algorithms, and accompanying conceptions of reality are in continual flux, while being in tension with one another too.

We are always encountering localized arrangements of these elements. The contents of news media vary in response to a broader cultural climate, but also in relationship to the network of technologies within which they are composed. Data structures for accessing news online vary as well: some news is delivered as text-based HTML; other news is broadcast as streaming video; and many sites now offer affordances for interaction and commentary. Meanwhile, today's social media, enabled by large scale databases, are increasingly the settings in which news is read, watched, annotated, and shared. Finally, algorithms for NLP and an awareness of their limitations have affected all these other elements: social media platforms use such algorithms to curate news content and target advertising toward readers; data structures include tags and other forms of annotation meant to ease the burden on NLP; and even the way that the news is made is poised to change, to be more structured from investigation through delivery. Someday soon, each news story may be conceived of as an algorithmically manipulable composite of claims and evidence, organized for not only reference but also recomposition and reuse.[68]

There are other, darker implications to these new configurations of media, data, databases, and algorithms. First, without low-cost algorithmically generated advertising that can zero in on particular demographics, respond to those audiences dynamically, and direct them to an immediate point of sale, most small local news outlets have closed. This has left local governments and communities virtually unsupervised; traditionally, local news has played an important watchdog role.[69] Second, the large national and international news outlets that have survived are considering how they might further conform to new expectations for everything to be searchable as data. Third, news-viewing habits are becoming more siloed. In 2019, we have come to accept that social media filter our news into personal feeds, effectively cutting off segments of the population from news that they might find disagreeable.[70]

Lastly, new forms of fake news have emerged. Although news-like propaganda and overzealous editorials have been around for a long time, networks of technologies that support the emergence of news data open up the possibility for more insidious ways to fake the news. In social media feeds, fake news propagates, indistinguishable from less biased, more conventional alternatives. The limitations of current instantiations of NLP that prevent its algorithms from accurately detecting fake news might be overcome someday by incremental technical improvements. But it is more likely that all the elements of news datafication will converge in new ways: news media will retool themselves, data formats will be restructured, databases and platforms for hosting the news will be reinvented, and the boundaries between what is real and what is fake will shift.

As in any arms race, fake news will continue to defy the algorithms devised to detect it. All these changes are happening in parallel but informed by one another. Understanding the locality of datafication means acknowledging all these shifting elements: reality, media, data, databases, and algorithms along with their interrelationships.

CONCLUSION

Data and algorithms are not stand-alone elements of computing, independently applicable anywhere. They are created in close coordination with one another, and with prevailing conceptions of the reality that they seek to represent, analyze, or predict. Moreover, they only function symbiotically, in contingent local conditions that are both materially and historically grounded. The news is a crucial site in which those contingencies might be better understood.

The nascent algorithmic turn in the analysis of news heralds several profound implications. First, old news is no longer trash; it becomes part of the consultable archive of what has been said. News archives and the algorithms that bring them to life can help us hold public figures and processes accountable, and protect against biases or even fake news. Yet the form of the news is not static. Producers of the news must choose whether to lean into or challenge the operation of normative algorithms for NLP. Already some news organizations are actively rethinking the way that news is generated, from robot reporters on the beat in areas like sports and weather, to more structured approaches to journalistic investigation that promise to make news data more accountable and accessible to remixing. In the coming years, all these elements of the news will continue to change and reciprocally inform one another: media, data, databases, algorithms, and even conceptions of what counts as news.

Beyond the roles that they play in the news, algorithms for NLP are pervasive: on social media, in internet searches, and even under the hood of the word processor I am using to write this book. I have barely begun to explore the multiple localities of NLP. Further research is needed to understand how such algorithms and the human labor that they subsume work together with data to shape important domains of public life.

The next chapter will continue to investigate algorithms as part of another computational structure: the interface. Already an element of everyday life, the interface also faces challenges posed by the locality of data. Rather than preserving the originating context of data, many interfaces today seek to recontextualize data, with critical implications for how data are made actionable and for whom. The illustrative case for that chapter is housing data, in many ways the highest-stake case in the book as well as the last one.

5

MARKET, PLACE, INTERFACE

PARCELID, TAXYEAR, ADDRESS, ADDRNUMBER, ADDRPREDIR, ADDRSTREET, ADDRSUFFIX, ADDRPOSDIR, ADDRUNTTYP, ADDRUNIT, OWNER, OWNERADDR1, OWNERADDR2, TAXDIST, TOTASSESS, LANDASSESS, IMPRASSESS, TOTAPPR, LANDAPPR, IMPRAPPR, LUCODE, CLASSCODE, LIVUNITS, LANDACRES, NBRHOOD, SUBDIV, SUBDIVNUM, SUBDIVLOT, SUBDIVBLCK

14 004600061108, 2011, 556 JOHN WESLEY DOBBS AVE SE,556,, JOHN WESLEY DOBBS, AVE, SE,,, MC LEAN DAVID, 556 JOHN WESLEY DOBBS AVE NE, ATLANTA GA 30312, 1638, 05W, 97360, 10800, 86560, 243400, 27000, 216400, 102, R3, 2, 0.172107, 14362,,,,

***Source*: Fulton County Board of Tax Assessors**[1]

DATA IN THE CULTURE OF REAL ESTATE

I live on the Eastside of Atlanta, Georgia, where I cannot go a single block without seeing a "for sale" sign. Many of the single-family bungalows on the market in this area are gut rehabs, bought on the cheap and flipped for a profit by independent speculators or major investment firms on Wall Street.[2] These homes are practically new—all but for the most foundational elements, a footprint or facade, which are required to preserve generous zoning laws concerning the size and placement of the house. On abundant empty lots or at the site of teardowns, newly constructed houses (often two or more times the size of surrounding dwellings) are springing up, outfitted and priced to sell to a new affluent population flooding the "Intown" neighborhoods of the city.[3]

Frequently, like me, these newcomers hail from outside the state. Meanwhile, traces of former communities linger in view. At the main intersection near my house is an abandoned bodega with a sign that reads "IFFY GROCERY"; the "J" is missing. An *Ailanthus altissima*—the "feral tree" from chapter 2, often characterized as a symbol of a neighborhood in distress—grows at the edge of the store's parking lot.[4] One block away, a yard is spotted with the personal effects of an evicted family that I never met: a laminate wood dresser, box spring, pair of jeans, and grade school activity sheet. But such sights are fleeting; the emptied-out house does not go unoccupied for long.

This latest housing shift, a rebound from the 2007 crisis, is made possible by many things: the changing lifestyles of well-heeled professionals who now put a premium on urban living, increasing economic inequality, subprime lending practices, and as I will focus on here, a "culture of real estate" enabled by the recent widespread availability of data on property values.[5] Communication scholar Joshua Hanan explains this emergent culture as one that combines nostalgic desires for domestic comfort with aspirations for profit and social ascendance.

Like many others purchasing a home in the area, my partner and I used a traditional real estate agent, but also the digital listings of available homes on platforms such as Zillow.com, which offers prices and inviting depictions of interiors, yet little indication as to the history of neighborhoods, or the implications of buying a house in the current cultural and economic climate. As in many urban centers across the United States, home prices within Intown Atlanta have fluctuated wildly over the past few years. In 2011, the median home value in Atlanta was $205,000. It dropped to $152,500 in 2012. By 2016, it had risen again to $250,000. But the cost of homeownership is not the only story. Between 2012 and 2014, 95 percent of rental units constructed in Atlanta were luxury apartments.[6] At the same time, affordable rentals are being demolished systematically to make room for units with a higher return for developers and landlords.[7]

5.1

A view of downtown Atlanta from the Crowne Plaza hotel. Image by the author.

When seen simply as a stream of incoming sales data, Atlanta can look like a city rising up or population left behind, depending on your point of view (figure 5.1). Either way, the future of Atlanta is increasingly viewed through data. Consumers most often encounter these data through what experimental humanities scholar Ed Finn writes about as an "interface layer," formed by tightly curated user experiences meant to shield audiences from the messy sociotechnical conditions of data collection as well as the implications of their use.[8] If data can be considered as texts, as I first suggested in chapter 1, interfaces are contexts: the settings in which data are meant to be fully understood.

This brings me to the fifth principle that supports this book's overarching claim: *interfaces recontextualize data*. Running counter to the lessons of previous chapters, today's interfaces often manifest the aspirations of digital universalism, introduced at the beginning of this book. Universalizing interfaces to data seek to further the ideology of placelessness by integrating data from anywhere and aiming to work equally well everywhere. In order to create and maintain this illusion, such interfaces first delocalize existing data sets, removing all traces of the places in which they are made, managed, and otherwise put to use. Then they present uprooted data within new contexts: unimpeded by the details of data production, unburdened by ethical quandaries that might accompany their use, and free from concerns about their unintended consequences. Such interfaces are known by user experience designers as being "frictionless."[9]

Whether you are looking for somewhere to live, a good meal, information about events in your area, or a ride to work, a new economy of interfaces stands ready to serve you through a series of transactions with data that can be carried out on any networked personal computing device.[10] The data that enable these services are created at the local level, collected by civic institutions or crowdsourced from the users themselves. They are rapidly mobilized by data brokers, who build and maintain national- or international-scale data infrastructures for profit.[11] The boosters of this new "smart" lifestyle are ushering in a new kind of individualism tailored for affluent and tech-savvy urban dwellers.

Consider their tag lines: Yelp, an online directory of restaurants, shopping, and other services, can make sure you "connect with great local businesses."[12] Nextdoor, a place-based social media platform, invites you to "discover your neighborhood."[13] Uber, a networked car service, equates "getting there" with personal freedom: "your day belongs to you."[14] Zillow, the real estate website, will help you "find your way home."[15] These interfaces promise not only access to data but also the operational context to easily act on them.

By operational context, I mean an interface that is procedurally generated from computer code, and composed of visual, discursive, and algorithmic processes that connect existing data to concepts as well as resources that can support their use. Indeed, interfaces are not places per se. Rather, as media theorist Alexander Galloway notes, interfaces are best understood as processes.[16] Visual processes, such as mapping or graphing, help users see patterns in data. Discursive processes offer ready-made narratives through which to frame those patterns as reality. Algorithmic processes enrich data by generating new value from existing inputs. These interface elements are more than representational because they transform data on the city—for instance, in terms of prices, distances, and rankings—into drivers for local and highly personalized behavior. As in the example of the arboretum in chapter 2, data don't just describe contemporary places; they are a functional part of the way that those places work.

This chapter brings a local perspective to the question, What are interfaces to data? This entails an analysis of the various processes that shape the way we encounter data in applications, such as those described above. The question of the interface is of deep relevance for anyone who wishes to think critically about systems that mediate relationships between people and data. On the face of it, designing an interface is a pragmatic problem of supporting data use. But it is also a problem with important social and even political consequences. Who can use data? How can they use data? And what can they use data for? Interfaces establish the subject positions that users of data are expected to adopt.

In the case of Zillow, which I will focus on here, you can be a prospective home buyer, renter, seller, or real estate professional. No resident, according to Zillow, is outside the market. But Zillow is not the only setting in which we might understand housing data. The potential interfaces for data are always multiple, enabling different forms of engagement and interpretation, with implications for what data appear to say.

The values of homes in the United States and other countries where property is on the market have long been determined in large part by context. The perceived worth of a home is not determined solely based on its age, square footage, or the number of bedrooms and bathrooms it contains. Home values fluctuate based on comparable sales in the area, changes in the neighborhood itself, interest rates, and even the time of year. What counts as context when it comes to pricing a home? The seller and buyer are the ultimate arbiters of that. Yet professionals—realtors, lenders, researchers, developers, and more recently, information technologists and designers—seek to influence perceptions of context by sellers and buyers of housing.

Today, context for home value is increasingly assessed through data. Although the housing crisis of 2007 raised important questions about the way we finance housing in the United States, it has failed to raise parallel and necessary questions about the way we use housing data. I intend to address some of those questions here, asking, How are housing data presented by commercial interfaces, and how do those interfaces shape perception and action in public life? Zillow takes input from public and private sources, such as tax assessments and sales records, in nearly every municipality in the United States. It uses these resources to shape, as much as any other commercial entity, the context in which nonexperts understand housing. Through a combination of visual, discursive, and algorithmic processes, Zillow demonstrates a range of ways for recontextualizing data. Moreover, I will show that the frames through which we examine such data have serious implications for the future of affordable housing. For this reason, it is necessary to reconsider the settings through which we look at, talk about, and calculate value with housing data.

Before I delve into the specific elements of the Zillow interface, however, I would like to examine what context means in relationship to data. Although the term is widely used in both academic and popular writing, its relationship to data is still being worked out.[17] My use of the term differs substantially from other uses prevalent in the study and design of information systems.

MODELS OF CONTEXT

"Let the data speak for itself": this advice comes from my colleagues in the field of information visualization (often using the term *data* in the singular, to my chagrin). Their suggestion concerning the design of interfaces to data seems responsible and even respectful. It honors and personifies the data. The sentiment is one that I hear everywhere in academia.[18] It is also prevalent in industry and public policy.[19] But I believe that the statement deserves a degree of scrutiny.

What does it mean for data to speak on their own behalf? Like other technological platitudes, the phrase contains several unexamined implications.[20] First, the statement treats data as an autonomous participant in conversation with humans. Second, it indicates that what data say is self-evident, requiring no interpretation. Third and more subtly, it suggests that data are currently marginalized and only need to be given an opportunity to speak.

And yet these conceptions of data do not fit with our everyday experiences. We see data using sophisticated visualization tools (that is, maps, timelines, graphs, and charts). Data are framed by a variety of discourses. And finally, data today hold a privileged place within contemporary scientific, business, and policy deliberations.[21] Indeed, imagining that data need to speak for themselves requires ignoring the various contexts in which we encounter data.

In response, some social researchers encourage us to "put data into context."[22] In particular, social scientists argue that the right context is crucial to understanding big data, which has frequently been uprooted from different places and times.[23] But this advice can be unclear as well. The commonplace definition of *context* in the *Merriam-Webster* dictionary is "the circumstances that form the setting for an event, statement, or idea, and in terms of which it can be fully understood." Yet this does little to illuminate how we might determine the right context for data. I contend that context isn't self-evident or ready-made; it must be assembled through practices grounded in preexisting knowledge systems. Context for data is operationalized in the form of an interface, focused on producing specific interpretations of data.

My operational perspective on context differs from the dominant modes by which context is accounted for in information systems.[24] Paul Dourish, a theorist of human-computer interaction, sums up the prevailing views in his article "What We Talk about When We Talk about Context." Although he writes specifically about definitions of context in ubiquitous computing, an area of research that explores the potential for computers to be distributed throughout the range of human environments, these same uses of the word can be found in discussions of data. Dourish juxtaposes a "representational" model of context, pursued by the majority of researchers in computing, with an "interactional" model, grounded in phenomenological inquiry, an area of philosophical thought centered on understanding individual human experience.

Dourish explains that in the representational model, context is a form of information or data, which is easily delineated, stabilized, and separated from the subject itself. Representational context is "a set of features of the environment surrounding generic activities ... [that] can be encoded and made available to a software system."[25]

Meanwhile, in an interactional model, writes Dourish, "context isn't something that describes a setting. It is something that people do."[26] As such, the context of any event or object can vary enormously depending on whom you talk to and when. Interactional context is relative, dynamic, spontaneous, and arising from activity. This differs from representational context, which is objective, static, and independent of interpretation. In order to avoid further confusion over the term, Dourish suggests that technologists leave aside the notion of context altogether. Instead, why not think about *practices* as the forming the settings for human interactions with computers?

That is exactly what I intend to do. Contextual practices, though, are not something that emerge spontaneously in an unselfconscious moment. Rather, such practices should be understood within culturally embedded knowledge systems, composed of inherited roles, concepts, and technological affordances.[27] In an interface, the practices that give data context are often codified as processes: visual, discursive, or algorithmic. Although an interface does not determine the way that data are used, it provides a procedural setting that shapes the roles and ways of knowing available to users. Clifford Geertz puts the relationship between context and culture as follows: "Culture is not a power, something to which social events, behaviors, institutions, or processes can be causally attributed; it is a context, something within which they can be intelligibly—that is, thickly—described."[28]

As mechanisms that establish the context for data, interfaces might be thought of as what Finn terms *cultural machines*—established and maintained through processes that are consciously designed to secure the value data and benefit particular audiences.[29] A dispute over the context of data is a disagreement over their meaning, but also their use: Who can access the data and to what end? In order to make sense of these three models of context—representational, interactional, and operational—consider a thought experiment used by Geertz.[30]

Three students are engaged in a seemingly identical action: momentarily contracting their eyelids on one side. For the first student, the action is involuntary—a twitch. A second takes an intentional action in response to the first, thinking that they are returning a wink. Finding humor in this exchange, a third student parodies the wink. Now suppose we only know about these events through data that document each eyelid contraction in the same way. Only by identifying the context—the knowledge system in which this series of actions unfolds—can we hope to understand the meaning of the eyelid contraction data: that they represent a twitch, wink, parody of the wink, and even undocumented cases of wink abstention.

Each of the various models of context introduced above would seek to put the data in context differently. A representational approach might take all available data describing the setting, such as geolocation, temperature, arrangement of students, and time, as the context for any single eyelid contraction. In contrast, following an interactional model would mean questioning whether these preexisting details are relevant. More likely, an interactional approach would entail supplementing the eyelid contraction data with accounts of context directly from the students who were there, perhaps from interviews. But looking for the operational context means something more: learning about the encompassing student culture in which various eyelid contractions and noncontractions take on meaning. After all, these students didn't invent the concept of winking.

An inquirer interested in operationalizing the data on winks would investigate structural questions. What kind of body language do students use to communicate? And what sorts of messages does that language privilege? Moreover, how can it be made accessible to outside observers? Indeed, defining the context in operational terms has a goal: it would allow one to not only analyze a series of eyelid contractions but also participate in their exchange and even take appropriate action (that is, suggest that the student with a twitch go to the doctor, and direct the one parodying to student theater). Although Geertz does not use the term *operational* in his assessment of this imagined event, his analysis falls along similar lines. He once again explains what culture means, yet this time in terms of eyelid contractions: "A stratified hierarchy of meaningful structures in terms of which twitches, winks, fake-winks, parodies, rehearsals of parodies are produced, perceived, and interpreted, and without which they would not ... in fact exist, no matter what anyone did or didn't do with his eyelids."[31]

In short, an operational context for data is a culturally defined setting in which participants are equipped with the resources and subject roles necessary to access, interpret, and take action on predetermined objects of attention. Although to put it that way is to suggest that cultural context is something settled and uncontested. That is not the case. Contexts that operationalize data are always under construction. Furthermore, disputes over context are common, sometimes with striking significance, as the case of Zillow reveals.

The design of interfaces is more than a means of communicating wink data. In a domain like housing, interfaces can have the highest stakes. Thus we must ask, What does an interface enable? For design is not just an issue of knowledge but of use too. We should consider how interfaces are rooted in normative cultural assumptions about what data can and should do. Data don't speak for themselves, any more than an eyelid contraction does.

In Zillow, data are processed for the user in three distinct ways: visual, discursive, and algorithmic. The visual context of Zillow is defined by the functionality of its map. Placing data on a map enables comparative reasoning, but only about things that have been given a geographic dimension. Zillow's main discursive context is that of "public data." Data in the public realm are increasing accessible, but at what cost? Finally, Zillow uses an automated valuation model, the "Zestimate," to further contextualize data. This algorithmic context offers an interpretation of data, but through a set of opaque and speculative rules. These three dimensions of interface are less about establishing the (capital *T*) truth of data on property values than about creating traction with users.

THE ANATOMY OF A FRICTIONLESS INTERFACE

> Shopping for a home was like being in a dark room where only the agent was holding a flashlight. She'd shine it on two or three homes—listings or "comps" she had chosen for you—but all you wanted to do was grab the flashlight and wield it yourself. Or, better still, just flip on the darned light switch to see it all. That's why we created Zillow: to turn on the lights and bring transparency to one of our country's largest and most opaque industries.[32]

Zillow is a leading online real estate marketplace seeking to redefine the context in which we understand housing by creating access to and avenues for action on data. The name is a portmanteau created by combining the words *zillion* and *pillow* (where you rest your head).[33] Zillow was founded in 2006 by Spencer Rascoff and Stan Humphries with the goal of estimating the value of every home in the United States. It is not a licensed real estate firm, which would require the company to submit to licensing rules and regulations in every state where it practices.[34] It has strategically intervened into the real estate market, however, in a way that has changed the work of many realtors and other professionals in the industry.

Trent, the Intown real estate agent who helped my partner and me find a house, confronts an uncertain outlook for his job. How can he continue to justify the cost of his services (commission in Metro Atlanta is typically 6 percent) at a time when almost anyone can access listings for sale and rent online? "I can't hold data hostage," he jokes. But Trent's situation is serious—one he equates with the circumstance of the travel agent a few decades ago. Orbitz, Travelocity, and Expedia, among others, have all but put an end to that vocation. "In the past, someone needed my services," Trent recalls of his early days in the business just ten years ago. "Buyers and sellers wouldn't know what houses were on the market without agents." Today Trent must find leverage elsewhere. It is no longer access to data that realtors provide, he argues. Rather, it is context: "the context necessary to understand what it might be like to actually live in a neighborhood or an apartment complex." From Trent's perspective, new points of access to data are not going away. Yet local agents will try to market their own understanding of context for data—one that is intentionally juxtaposed to Zillow's ten-thousand-foot view from above.

Zillow is not the first web company to tread into real estate.[35] Moreover, its way of operationalizing data is not original or unique. It is just one of the many data brokers that seek to produce surplus value from available data on housing.[36] Still, Zillow's recent purchase of Trulia—another major platform for home listings focused more on user experience than analytics—has consolidated its position as a market leader in the United States.

As a representative explained during a routine webinar I attended to learn more about Zillow, the company's unofficial motto is "data wants to be free." This may seem a laudable, emancipatory goal. In addition to echoing the earlier sentiment "Let the data speak for itself," it recalls the famous statement "information wants to be free," expressed by the iconoclast Stewart Brand.[37] But data are never free, only recontextualized. Zillow presents data in a new setting defined by the processes underlying its interface: visual, discursive, and algorithmic. Indeed, Zillow offers important lessons on how to put data in context. Its approach should give us pause, though, for it demonstrates that interfaces are not neutral.

Zillow is invested in furthering the culture of real estate by creating a seemingly rational, economic setting in which individuals are given access to information and encouraged to make choices based on their own self-interests. The effects of this setting are damaging in ways that Zillow obscures. For although users may believe that they are independent actors, the demand they place on the market works to increase the value of all property in their area and limit the availability of affordable options. As I will show, Zillow not only supports a market-based approach to property but also works to increase anxiety among its users by emphasizing instability in the market with its Zestimate algorithm. Let us unpack all three dimensions of Zillow's interface to understand how they are constructed, and their effects on both data and housing.

The Visual Dimension

For Zillow, putting data in context starts with positioning them on a uniform map (figure 5.2). Properties for sale, for rent, or otherwise of interest (that is, foreclosure or a category simply labeled "make me sell") appear as colored dots on a faint, gray background showing a network of streets, parks, bodies of water, and place-names.

Hovering over a dot brings up a tag including a small thumbnail image, price, number of bedrooms and bathrooms, and square footage. This intentionally generic setting—the same everywhere across the geography where Zillow lists properties—frames our understanding of housing data not by showing the conditions of their production (representational context) or how they might have been used in the past (interactional context) but by suggesting what can be done with them today. The map is operational.

A venerable technology for visual reasoning, maps are recognizable and accessible to most of Zillow's users.[38] They offer a structure for making sense of data through spatial patterns; they show where the listings are located. This enables comparative readings of listings (that is, "these listings are close to one another") as well as readings of each data point within a matrix of surrounding features (that is, "these listings are close to a park"). These are operational relationships, and as such they can serve as the basis for consumer decisions about real estate—a domain in which it is said that the three most important indicators of value are location, location, and location.[39] The map does not merely register the locations of real estate in the real world.[40] Rather, it produces a reading of location using a narrow set of visible relationships (to, for example, select streets, bucolic parks, and highly ranked schools). In this way, the map participates in the production of reality for real estate by establishing or confirming conceptions about what conditions of location determine value.[41] Thus, putting data in the context of the map is not a retrospective practice. The map does not reunite data with some preexisting setting. Zillow's map is operational because it stimulates actionable interpretations of location and its implications for home value.

But maps do not "unfold" in isolation.[42] Zillow's map is framed by other media and modes of access to the underlying data (figure 5.3). Above the map is a search bar with filters for listing type, price, number of beds, and more that can be applied to further narrow the number of listings displayed. To the right of the map is a column of property images, mostly facades. Each is annotated with more details about individual listings such as the number of days it has been on Zillow, the name of the listing agent, and the type of sale (house for sale, preforeclosure, or lot/land for sale).[43] These images can act as links to a full-screen view of an individual listing (figure 5.4).

The additional elements of the Zillow interface serve to put the map itself in relief. They help users interpret the map as a collection of commodities: locations valued because of their potential to be bought and sold, not because of their historical significance as places or the circumstances of the people who currently live there. The visual elements of Zillow's interface illuminate a number of things: which data points matter, the relationship between the points, the meaning of the space in between them, and the connection between data and any secondary media.

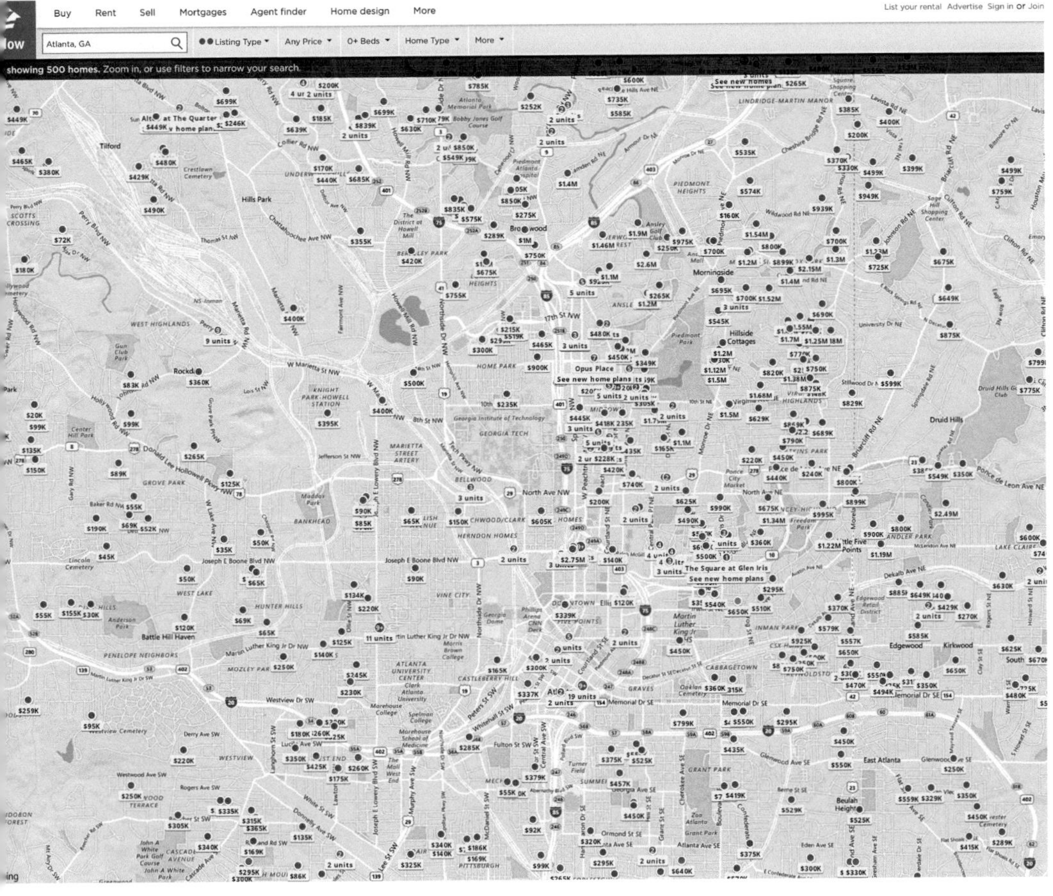

5.2

Intown Atlanta, populated by home listings, as depicted on Zillow's map.

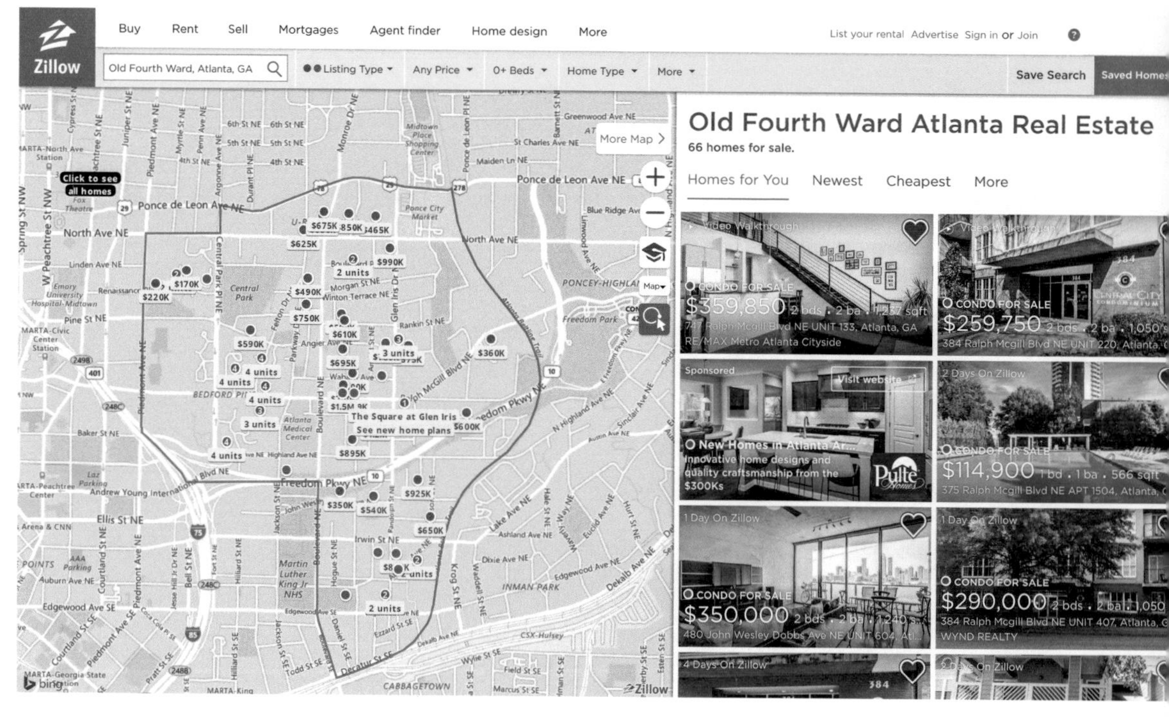

5.3

The Old Fourth Ward neighborhood of Atlanta as depicted on Zillow's map.

CONTACT AGENT
SAVE
SHARE
HIDE
MORE
EXPAND
CLOSE
Public View
Owner View
Old Fourth Ward
Irwin St NE
Boulevard NE
Lake Ave NE
Dekalb Ave NE
Austin Ave
bing
Zillow
658 Mcgruder St NE,
Atlanta, GA 30312
3 beds · 3 baths · 2,100 sqft
COMING SOON
$649,900
Zestimate: $649,025
Expected on market Apr 30 .
Contact now for more info
You could not find a better location than this charming 3BD/3BA bungalow in the Heart of Old 4th Ward. Set on a sunny corner lot just steps from Krog Street Market & the Beltline, the home boasts big rooms, light-filled spaces, high ceilings, and hardwood floors throughout. Updated chef's kitchen with granite countertops, SS appliances, and breakfast bar. Open Living and Dining Room with 2-story vaulted ceiling. Large master suite with built-in reading nook, skyline views, his & hers closets, stylish master bath & big office or sitting area. 2 additional guest bedrooms on the Main - one with ensuite bath and double French doors to balcony. Big front porch with views of the Water Tower. Private fenced side yard w/garden box and deck. Off street parking for 2 cars. Double pane windows, concrete siding, kitchen pantry, big laundry closet, two storage sheds, gleaming hardwood floors. Tons of open space and natural light. All this literally a few steps from the Beltline, Krog Market, Studioplex, & Stoveworks!
EST. MORTGAGE
$2,641/mo
Get pre-qualified
CONTACT AGENT
(39)
7 Recent sales
Your Name
Phone
Email
I am interested in 658 Mcgruder St NE, ATLANTA, GA 30312.
Contact Agent
By pressing Contact Agent, you agree that Zillow Group and real estate professionals may call/text you about your inquiry, which may involve use of automated means and prerecorded/artificial voices. You don't need to consent as a condition of buying any property, goods or services. Message/data rates may apply. You also agree to our Terms of Use.
Learn how to appear as the agent above

5.4

A single home listing from the Old Fourth Ward neighborhood of Atlanta, as seen on Zillow.

The Discursive Dimension

Beyond the setting established by Zillow's map, the company's interface also suggests the terms by which users should talk about housing data. These discourses serve to establish and stabilize Zillow's use of data as a legitimate representation of the world of real estate. Indeed, discourses define what can and cannot be comprehensibly said about a subject.[44] Among the most important discourses used by Zillow is that of *public data*.[45] In a list of frequently asked questions on its website, Zillow explains *public information* (a related but more expansive term than *public data*) as the way and reason that it knows things about your house: "Zillow receives information about property sales from the municipal office responsible for recording real estate transactions in your area. The information we provide is public information, gathered from county records."[46]

The term *public* tells users that Zillow's map is based on data from an open and authoritative source: the municipal office in their area. This legitimizes both the data and Zillow's use of them. For although Zillow operates outside the boundaries of any particular municipality (remember it is not licensed anywhere), it is making fair use of data created by and for the people. Moreover, invoking the public context of these data protects Zillow from requests by homeowners to have them removed from the site. After all, these are not private data. And though—as the reader might come to expect by now—Zillow acknowledges that municipally created data may contain local contingencies, such as skewing effects or errors, the company takes no responsibility for those. It is up to homeowners to show proof of anomalies that might affect public perceptions of the value of their own house. Ironically, all these assertions about Zillow's rights with respect to the data are wrapped in the language of public empowerment: "Our mission is to empower consumers with information and tools to make smart decisions about homes, real estate, and mortgages. For this reason, we do not remove public record property data from Zillow unless it is shown to be erroneous."[47]

Beyond its discourses on public sources, Zillow also cultivates a perception of its map as a virtual public space and invites contributions of data from private sources to be made broadly accessible. There are two ways that this can happen: realtors can contribute their own listings—and pay a fee to have their profiles promoted in association with those listings—or owners can contribute "house facts" in order to improve the online image of their property. As one real estate technologist remarks, "If you are able to give people a real-time value of their home, they are going to check that value and ask: what can we do to update that value? The Zestimate [Zillow's algorithm for predicting property values] is a powerful consumer engagement instrument."[48] By making its database open to public reading as well as public writing, Zillow fashions itself as the "Wikipedia of housing": a democratic, free, and transparent context for sharing data publicly.[49]

In tension with the discourses on public data and public space is that of a personal journey. "Find your way home" is the welcoming message on the Zillow front page.

"You are in the driver's seat." The implication is that Zillow is a vehicle that can be used in the journey toward homeownership. This second discourse positions the platform as a navigational aid in an individualized search for *home*—a term used as a synecdoche for personal comfort, security, and belonging—through a bewildering landscape of consumer options. While the space of data on Zillow is public, the journey through that space is private and the implication is that it should be guided by individualized interests as opposed to the public good. In this way, the culture of real estate is strengthened.

The result is the creation of a public setting in which—following the flawed logic of digital universalism outlined at the beginning of this book—everyone has access, but no one is equal. Users come to the map with different resources for buying, selling, or renting, and Zillow lets them know right where that places them. Thus, the discursive elements of the interface define the relationships between people and data: who owns them, manages them, or uses them, and who doesn't, and what stories about the nature of the data justify these attachments and exclusions.

The map, its media annotations, and an overarching discursive framing put Zillow's data in context—one that isn't a reconstruction of the origins of data but rather an operational setting that makes the data actionable. Yet the interface marshaled by Zillow extends beyond these visual and discursive elements. Zillow has been successful in large part because of an additional computational layer of context that it brings to the existing set of housing listings.

The Algorithmic Dimension

Zillow's "rules of real estate" establish the final dimension of its interface that I will discuss.[50] The company not only accumulates data from a variety of sources but also extracts a surplus value from those data in the form of computationally generated predictions. Using the data that Zillow has assembled on sales and historic valuations of homes in a particular area, the company produces estimated values for properties, including many that are not currently on the market. This process of triangulating property values is called an automated valuation model. The outputs of Zillow's model, comically dubbed Zestimates, are generated for nearly every home in the United States.[51] More generally, the algorithmic dimension of an interface is any procedure that enriches data by generating new values based on existing inputs.

At the time of this writing, a Zestimate was calculated for about a hundred million homes nationwide using public data as well as information contributed by realtors or homeowners.[52] The physical characteristics of a home (i.e., its location, square footage, and number of bedrooms) and its past sale prices, as well as the prices of comparable homes nearby, are analyzed using proprietary valuation rules. Instead of relying on a single complex model of the entire US market, Zillow depends on simpler, albeit obfuscated, localized models (sometimes at the scale of a single street) to account for different market situations.[53]

CHAPTER 5

And yet the Zestimate still struggles to account for the contingencies of the data on which it is reliant, such as the politics of tax assessments, dated house details, fake sales, and even false sale prices that are a part of many purchasing negotiations. The Zestimate makes up for its limitations by being dynamic. Home values are discarded every night and built again in the morning using fresh data that incorporate changing conditions.

Highlighting the context that the company brings to existing data, Rascoff and Humphries, the founders of Zillow, argue that it is not the data but instead the Zestimates it extracts from those data that differentiates it from other online real estate sites. The founders claim their Zestimate has a pulse on market conditions in local contexts across the country, and furthermore, that it can make accurate predictions on where the market is headed in the near future. For Zillow, the future is just another discursive element to be marshaled in operationalizing housing data.

The Zestimate is an algorithmic system, explained by anthropologist Nick Seaver as an "intricate, dynamic arrangement of people and code."[54] Understanding it means examining both its technical details and the social practices of its creators. This effort is complicated further by the fact that, like most algorithms, the Zesimate is a closely guarded secret and requires local knowledge to be decoded. Fortunately, we can learn much from simply looking at how it is discursively framed.

Like the map, the Zestimate has its own discursive elements. Users are told, for instance, that it is not meant to replace realtors but rather to connect them (as well as a range of other real estate professionals) to potential buyers and sellers. Take as an example an excerpt on how to deal with the Zestimate written by the company on its site.[55] Zillow's literature invites users who are selling their homes to quickly look past the Zestimate if it doesn't "feel" as accurate and up-to-date as possible. Moving along, sellers should first (no surprise) update their Zestimate by making Zillow aware of improvements to the home that aren't reflected in the "home facts." Next, sellers are encouraged to check the "comps" that Zillow provides—comparable recent sales within the neighborhood in the last sixty days. The market can change rapidly, and a sale six months out may be a poor indicator of what a house is currently worth; this is just an indication of the incredible instability and unpredictability of these values. Then sellers are prompted to think "psychologically" about the sale, such as by using findings from consumer research on the way that "magic" numbers like $299,000 instead of $300,000 can stand out from a cluster of other prices in the same neighborhood. Finally, explains the site, a number of other seemingly external factors might affect your home price including the time of year, interest rates, and job market in the area. Sellers who find all this overwhelming are encouraged to get a comparative market analysis from an agent or even a professional appraisal, which can run from $250 to$400.[56] At each step, sellers are encouraged to put the Zestimate itself into context using feelings, facts, comps, pop psychology, and the general economic climate.

Furthermore, the company talks about the need to understand the Zestimate as part of a social process. A recent public exchange between Zillow and a critic of the company exemplifies this. Realtor David Howell, the chief information officer at McEnearney Associates, calls into question the way that Zillow has intervened in the housing market in an article titled "How Accurate Is Zillow's Zestimate?" He notes that "on average, those 'Zestimates' are within 5 percent of the actual value of a home just half of the time." In other words, Zillow's estimates are "about as good as a coin flip."[57]

Responding to this challenge, Humphries, the "architect" of the Zestimate, weighs in on what he believes is absent from Howell's analysis: context. He says that "McEnearney disparages the fact that less than half of Zestimates are more than 5 percent off from sales prices as 'wildly inaccurate and inconsistent,' without much context as to how that level of accuracy compares to other opinions of value."[58]

Humphries goes on to argue that the accuracy of the Zestimate cannot be evaluated on its own terms—which is another way of saying that the data produced by the Zestimate don't speak for themselves. Rather, Humphries implores, we should understand the Zestimate's accuracy in relation to the accuracy of listing prices set by real estate agents.

> In our eyes, a helpful analogy here is WebMD, the large and popular online health resource. We've all searched online to research our ailments—"what are the symptoms of Strep Throat?"—but then we go to a doctor for a proper diagnosis. We need professionals to help us interpret and treat what anyone with an Internet connection can find in twelve seconds on Google. The doctor's role, then, is a little different, but it's definitely not diminished. The same is true of the real estate agent. Home purchases are infrequent, emotional, and expensive. High stakes command high expertise. But that doesn't mean people shouldn't be armed with the best possible information to help them navigate the process.[59]

Consider what Humphries is suggesting about the context of data analysis: the Zestimate, a computational model, should be understood in relation to other sources of data in a market in which no prediction can be trusted completely. Humphries acknowledges that the Zestimate is often incorrect. The prices set by real estate agents are more accurate. But not by as much as you'd expect, he says. There is a place for automated valuation models, suggests Humphries, as a "starting point for a conversation."[60]

The automated valuation model is not meant to replace realtors but, as I have said, connect them to buyers and sellers; this is Zillow's business strategy. It doesn't depend on the accuracy of its automated valuation model to make money. Zillow makes money on subscriptions from realtors and other professionals, who pay to advertise on the site. This ad-based model of revenue is not unlike those used by other web platforms that host social media including Facebook and Twitter.

In this exchange, we get a contextual explanation of the Zestimate. The model is good enough if it catalyzes a certain kind of social relation between real estate professions and potential clients. Again, we see that Zillow cares less about the truth of data than their tractability: the ease with which they can be used. Considering all this, realtors appear to be the real clients of Zillow—albeit not always happy ones.

Zillow has recently begun to err on the side of pricing homes too low. This has some obvious benefits.[61] For one, it may protect Zillow from accusations of further inflating the housing market. More important, though, a low price will induce anxiety on the part of sellers, making them more likely to, first, contribute additional "house facts" that might bring up the Zestimate, or second, turn to one of the options that Zillow offers for increasing home value: ad-laden advice columns, improvement projects (it can introduce you to a good contractor!), or a relationship with a subscription-paying realtor.

Algorithmic elements of Zillow's interface are a form of what Janet Murray and other digital media scholars call *procedurality*, "the computer's defining ability to execute a series of rules."[62] More specifically, the Zestimate might count as a type of what Ian Bogost labels "procedural rhetoric."[63] Bogost writes about how games can persuade through an expressive system rather than an explanation. To adapt this framework, we might say that the Zestimate, like a game, makes claims about how property values operate. It holds that not only square footage or the number of bathrooms and bedrooms but instead a host of other, sometimes specifically local characteristics are intertwined with value. We shouldn't see this as merely a representational system, as in the case of the persuasive games that Bogost studies. Rather, Zillow is an operational part of the way that the housing market works today. Zillow's rules form a system that homeowners can interact with by updating their home facts or simply checking their Zestimate regularly. Zestimates may be wrong. But it may not matter, if they lure users into a conversation with realtors and other professionals who pay Zillow to stay in business.[64]

In summary, the visual, discursive, and algorithmic dimensions of Zillow's interface work to delocalize data by establishing a new, seemingly generic context for their use. But everywhere Zillow is used, it is creating new local effects and subjects. When you peruse the Zestimates and imagine how you might afford your dream home, when you look for a fixer-upper that you can flip for an easy profit, when you click on the handy "make me sell" button, anxiously considering "how much could I get for my home today?" you become a desiring subject in the culture of real estate. These actions may seem minor. Yet they are incremental contributions to market inflation. And in a specific neighborhood, they can easily lead to speculation on local properties, increased prices, and ultimately the displacement of existing residents who cannot afford to stay. Existing low-income homeowners or renters, who fear being priced out of their local markets, are the invisible subjects of Zillow.

HOUSING DATA IN A CIVIC CONTEXT

The elements of Zillow's interface establish a consumer-oriented setting for thinking about housing data aligned with the culture of real estate. It is a setting in which users can compare housing options, contribute data on their own homes, and begin a conversation about what, based on data, can reliably be predicted about the market for housing. This context is enabling for many prospective buyers and sellers, not to mention the realtors and contractors who want to connect with them. But it is an interface rooted in knowledge systems that privilege self-interest and speculation, while not guarding against the accompanying potential for harm in localized places where its data are made or used.

Zillow's interface isn't the only possible way to recontextualize housing data. One way of discovering alternative contexts is by considering the existing cultures of housing to which Zillow does not connect.[65] As an interface focused on individual choice, it is not a setting in which to think about our collective responsibility to treat housing as a human right. Indeed, Zillow draws attention away from the broader impacts of the housing market.

I would characterize the agenda of Zillow's interface as *consumerist*. An alternative interface might use visual, discursive, or algorithmic elements in order to counter consumerist trends by questioning the property values as well as the social impacts generated by the market. I would call this a *civic* context, revolving around "the use of digital technologies to shape public life for the common good."[66]

Rather than positioning us as individuals empowered to act in a public market, a civic approach to housing data turns away from narrowly consumerist goals. As I will show, it can introduce the timeline as a visual element, gentrification as a discursive setting, and housing policy as an algorithmic frame. Such dimensions of context highlight the public impact of individual action and encourage us to care about the broader community in which we live. This civic context is not a better, more socially conscious version of Zillow. It is a corrective to Zillow; it is a countercontext that can help reshape how we think about and handle property value in the United States.

To illustrate the local implications of a civic context for housing data, I will focus on data from neighborhoods on the Eastside of Atlanta, where I live and also work with organizers and researchers who are fighting to preserve as well as expand low-income housing opportunities in Intown Atlanta. These housing advocates seek to understand how new forms of consciousness and new policies might mitigate the negative consequences of the rapid changes that we have witnessed—principally the displacement of low-income and predominantly black communities that have been living here for decades.[67] Atlanta's legacy as a "Black Mecca" is increasingly under threat.[68]

The dimensions of context introduced below frame housing as an issue that transcends its depiction in the market. Housing shapes the composition of our communities—racially, culturally, and economically—as well as access to resources, such as education, jobs, and transportation, not to mention a healthy living environment. Housing has a long and painful social history in Atlanta: from whites-only "park" neighborhoods built during Reconstruction, to civil rights era redlining policies that prevented people of color from acquiring mortgages, to the recent demolition of historically black neighborhoods on Atlanta's Westside so as to make way for hedonistic sports stadia.[69] Even today, the city is one of the most segregated and economically unequal in the United States.[70] Could an alternative interface, composed of visual, discursive, and algorithmic elements that enable civic operations on data, help reveal the downsides of the current trajectory of development in Atlanta?

Let us start with a different approach to visualizing housing data. Zillow offers the map as an accumulation of currently available purchasing options. In doing so, it overshadows the question of how that map came to look the way that it does. Visual settings need not be limited to geography, though. A simple timeline can reveal urban change quite dramatically. The images that follow demonstrate how housing data might be put into a visual context defined by temporal relationships.[71] Similar to Zillow's approach, these visualizations rely on publicly available data from the Fulton County tax assessor's office on the total appraisal value for single-family homes in the Old Fourth Ward, a historically black area of Atlanta where Dr. Martin Luther King Jr. lived and is now memorialized.[72] In contrast to Zillow's map, which is addressed to buyers and sellers, this image is for a different audience: a group of organizers in Atlanta collectively known as the Housing Justice League, working to put pressure on policy makers and raise awareness among the broader public. A grassroots, member-led organization, its mission is to "empower renters and homeowners to self-organize and defend their right to remain ... [and] fight to preserve affordable housing, prevent gentrification, and build neighborhood power for an Atlanta-wide housing justice movement.[73]

On Zillow's map, visual space represents geographic relationships. Properties are meant to be understood in terms of their location relative to one another as well as features of the surrounding area considered important for sale value, such as significant roads, parks, and bodies of water. In the images below, visual space is temporal. Thus, I use some of the same data harnessed by Zillow to tell the story of how the Old Fourth Ward has changed over six years from 2011 to 2015.

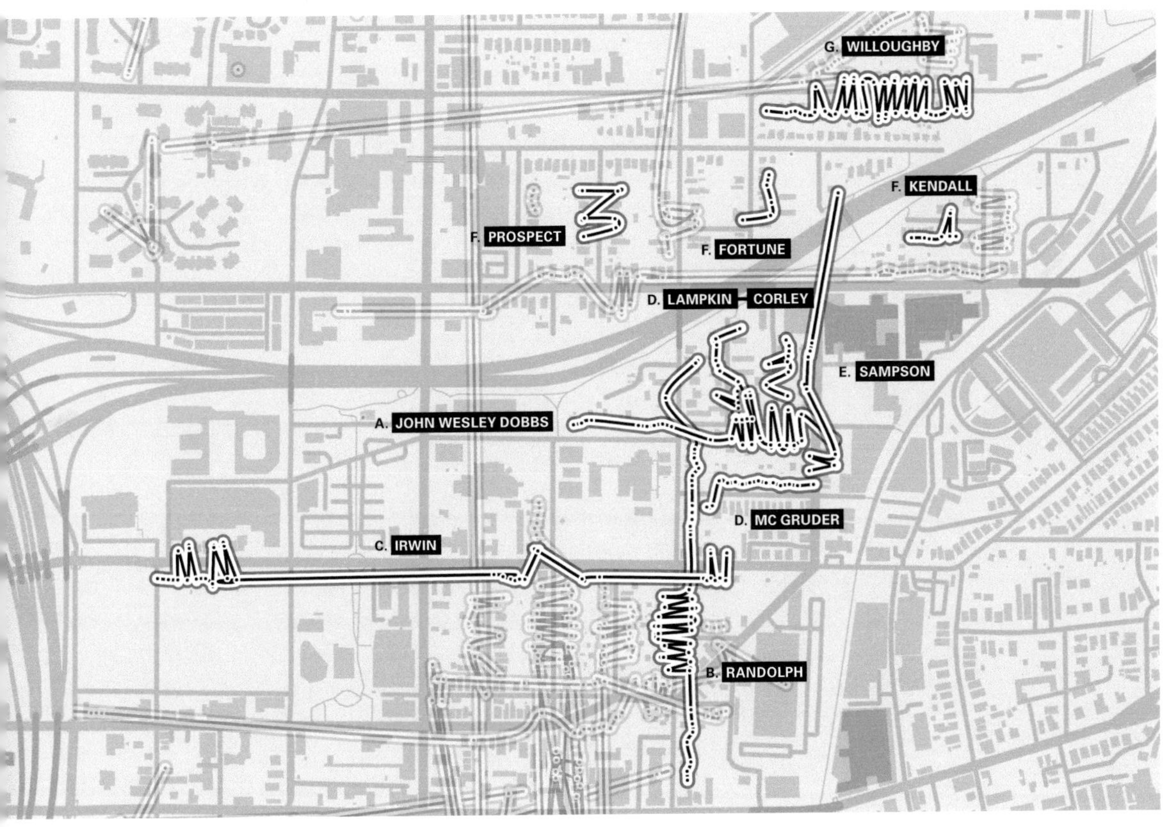

5.5

Map of streets in Atlanta's Old Fourth Ward used in property value timelines. Image by the author and Peter Polack.

540 — John Wesley Dobbs — 669

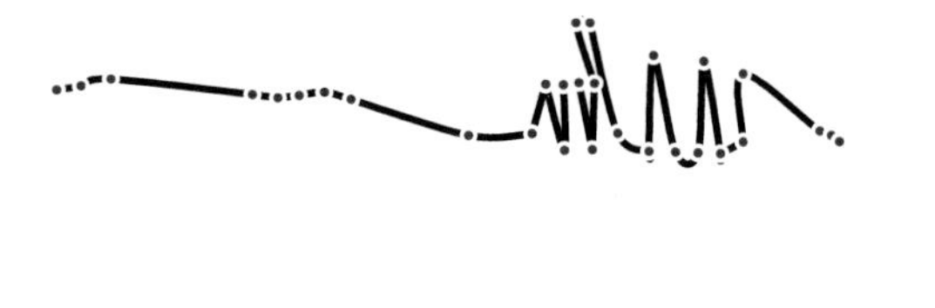

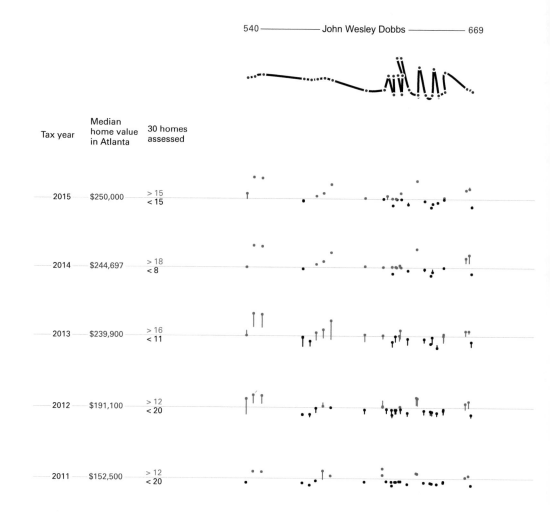

Tax year	Median home value in Atlanta	30 homes assessed
2015	$250,000	> 15 < 15
2014	$244,697	> 18 < 8
2013	$239,900	> 16 < 11
2012	$191,100	> 12 < 20
2011	$152,500	> 12 < 20

5.6

Property value timelines. Image by the author and Peter Polack.

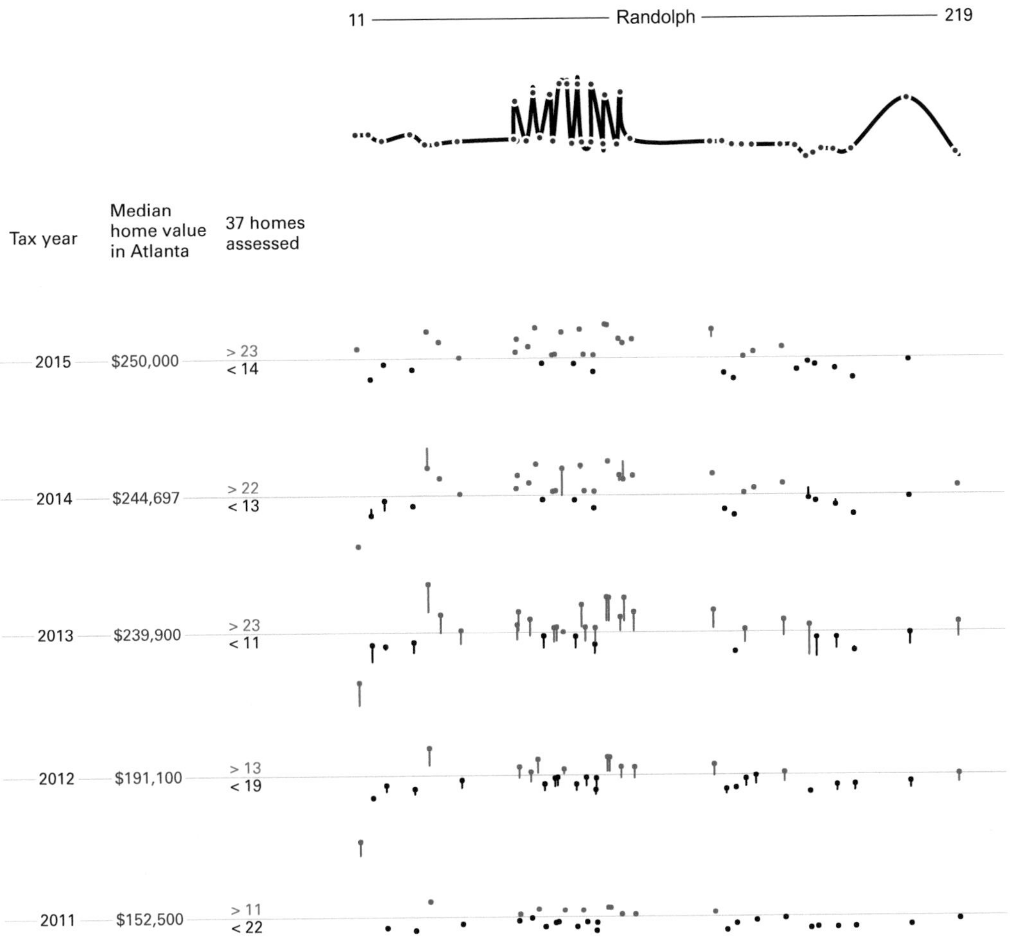

11 — Randolph — 219

Tax year	Median home value in Atlanta	37 homes assessed
2015	$250,000	> 23 < 14
2014	$244,697	> 22 < 13
2013	$239,900	> 23 < 11
2012	$191,100	> 13 < 19
2011	$152,500	> 11 < 22

341 —— Irwin —— 614

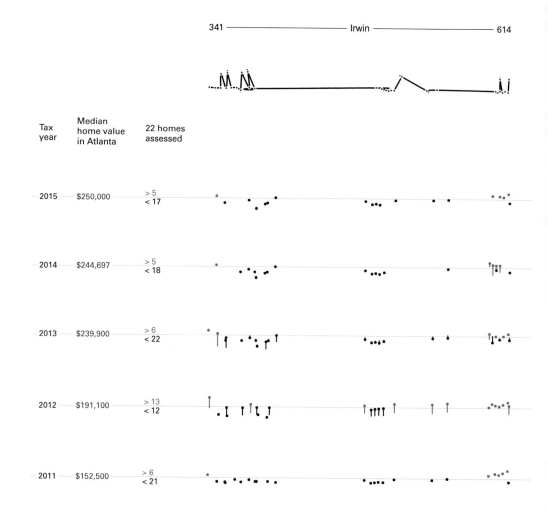

Tax year	Median home value in Atlanta	22 homes assessed
2015	$250,000	> 5 < 17
2014	$244,697	> 5 < 18
2013	$239,900	> 6 < 22
2012	$191,100	> 13 < 12
2011	$152,500	> 6 < 21

5.6 (continued)

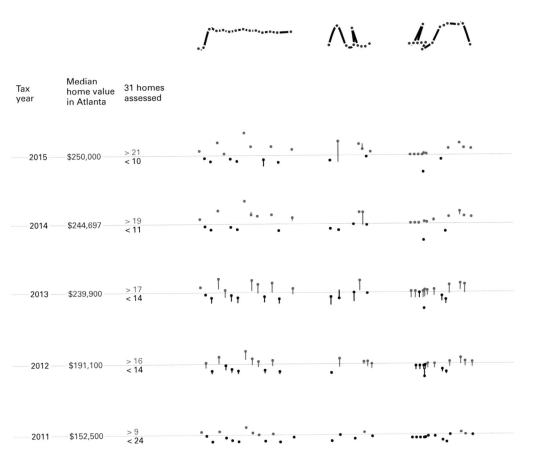

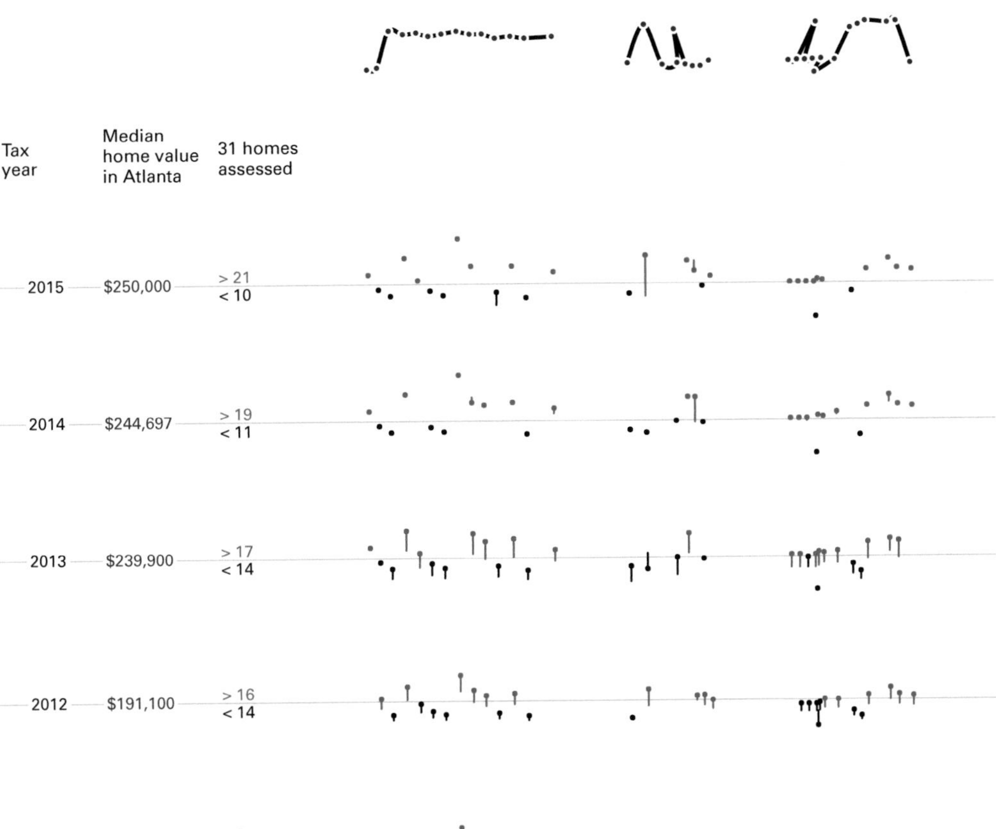
603 — Mc Gruder — 658
209-Corley-237
203-Lampkin-241
Tax year
Median home value in Atlanta
31 homes assessed
2015
$250,000
> 21
< 10
2014
$244,697
> 19
< 11
2013
$239,900
> 17
< 14
2012
$191,100
> 16
< 14
2011
$152,500
> 9
< 24

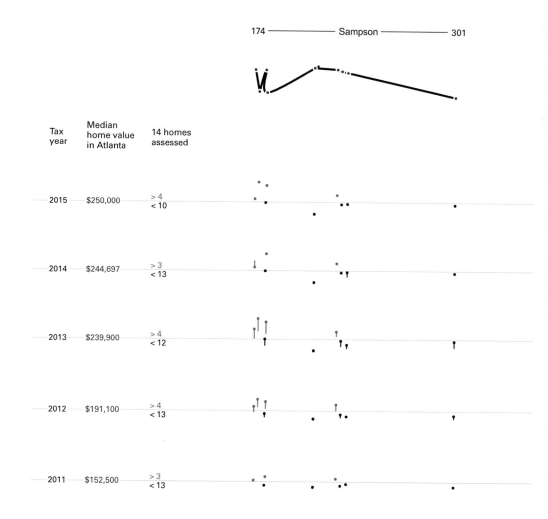

Tax year	Median home value in Atlanta	14 homes assessed
2015	$250,000	> 4 < 10
2014	$244,697	> 3 < 13
2013	$239,900	> 4 < 12
2012	$191,100	> 4 < 13
2011	$152,500	> 3 < 13

5.6 (continued)

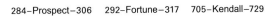

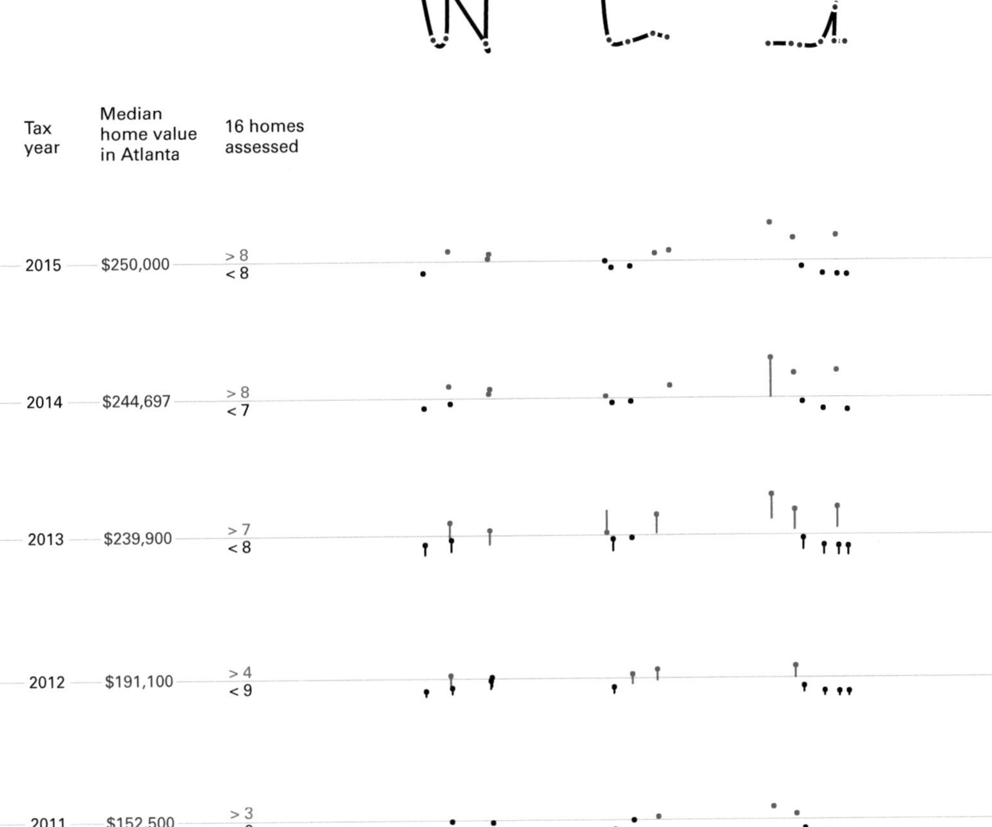
284–Prospect–306
292–Fortune–317
705–Kendall–729
Tax year
Median home value in Atlanta
16 homes assessed
2015
$250,000
> 8
< 8
2014
$244,697
> 8
< 7
2013
$239,900
> 7
< 8
2012
$191,100
> 4
< 9
2011
$152,500
> 3
< 9

637 —— Willoughby —— 736

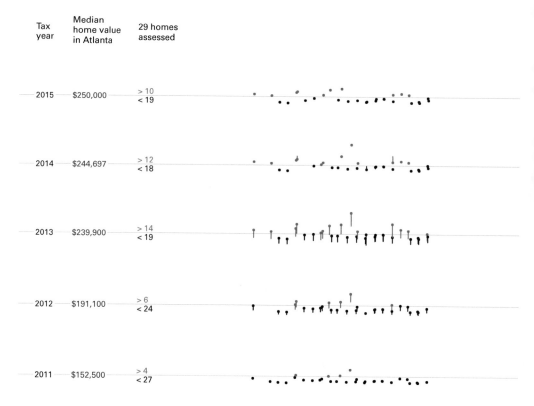

Tax year	Median home value in Atlanta	29 homes assessed
2015	$250,000	> 10 / < 19
2014	$244,697	> 12 / < 18
2013	$239,900	> 14 / < 19
2012	$191,100	> 6 / < 24
2011	$152,500	> 4 / < 27

5.6 (continued)

The preceding images are meant to be read as a grouping. The first is a map of the Old Fourth Ward. Here, property data from Fulton County tax assessments have been geolocated, then connected and labeled by street. The lines created by connecting each property do not follow the physical path of the streets. Instead, I think of each line as a signature for the street. These may look strange, but I've found them helpful as graphical references that can be easily used to identify and differentiate the streets. In subsequent images, each street is broken down into a set of graphs that track changes in the assessed value of individual properties. Because of their size and the way that they are juxtaposed with one another, such graphs are sometimes referred to as *small multiples*.[74]

Each small multiple displays all the properties on a single street and how they have changed in value from the previous year, according to tax assessment data.[75] An individual mark represents a single-family home, where the number of "living units" on the property is one. There are also multifamily homes and even a few large-scale apartment complexes in the Old Fourth Ward, but these don't appear on the graphs for the sake of simplicity. Instead, the graphs tell a subset of the larger story.

A node at the terminus of each mark indicates the price assessed during that tax year. The length of each mark represents the relative change in price since the previous year. Running horizontally across each graph is a thin gray line representing the median home value in the city of Atlanta during a single year. The median is the middle value in a list of all property sale prices in Atlanta, not just those in the Old Forth Ward (that is, value number fifty out of a hundred values).[76] Marks that surpass the horizontal line are properties valued above the median. Defined simply in terms of their market value, these are within the top half of all properties in the city. The marks are also color coded to reflect this division: red represents property values above the citywide median, and black represents property values below.

For example, there were thirty-seven properties assessed along Randolph Street in 2015 (figure 5.6). On the 2015 graph for Randolph, the distribution of property values is twenty-three above and fourteen below the city median. In this year, Randolph was in line with other markets across the city. But it wasn't always this way. The 2011 graph for the same street depicts a different pattern. There were only thirty-three properties; eleven properties were above the city median, and twenty-two were below. The additional four properties in 2015 are probably a result of parcel subdivision and new construction. These additions further indicate a growing market. Each figure in the overall set presents a different street in the Old Fourth Ward and the changes in assessments it has seen.

One more note about the locality of this data: the earliest assessment in these graphs is from 2011 because in that year housing prices finally began to stabilize after a period of wild fluctuation due to the 2007 housing crisis. In fact, the Fulton County assessor's office had a moratorium on assessments in 2008, 2009, and 2010 intended to hold values in place during the "great recession." The market value assessment in 2011 was the first in four years. The data set used above is just one example of the numerous county-level records that Zillow draws on in order to assemble its platform. The moratorium underscores a point, made throughout the book, about how data are shaped by local conditions. Only by understanding such data as a composite, created in disparate conditions, can it be leveraged responsibly.

Beyond conveying the local attributes of housing data, this exercise is meant to show that context matters. By taking home values out of a geographic setting and putting them into a temporal context, the images above open up new questions about possible operations that can be done with housing data—questions that start by recognizing that the Old Fourth Ward is a neighborhood in transition. What does it mean to purchase a house in this neighborhood? Who is moving in, and who is moving out?

In order to address these questions, I want to offer an alternative discursive setting for thinking about the rapid change in home values. *Gentrification* can be characterized by a variety of effects: rising home values, the cost and availability of local services, and the overall availability of low-cost housing.[77] Through the discourse of gentrification, rising home values in a neighborhood can be understood in terms of the displacement or marginalization of low-income residents, with consequences on their employment, relationships, and even health.[78] The new affluent members of the neighborhood—the "gentry" in gentrification—might invite high-priced restaurants and stores, inaccessible to the previously existing residents. Finally, a loss of affordable housing overall can be seen as a major cause of declines in social and economic diversity.[79]

Such changes are unfolding across Intown neighborhoods in Atlanta. Affluent home sellers are not unaware of these changes, and indeed stand to benefit from them. Their language, captured in the text of home listings on Zillow, often frames changing prices as an opportunity rather than a danger. Although these listings do not use the term *gentrification*, they gesture to related patterns in the market.[80] Some listings emphasize a newfound stability: "Excellent location in *established* neighborhood but with all the modern conveniences and finishes of a new home" (emphasis added). Others sell the transformation: "This authentic urban loft in *hot* Reynoldstown boasts soaring ceilings, concrete floors, brick walls, skylights & amazing open space!" (emphasis added).[81]

These discourses do not merely reflect market conditions; they are an operational component of how the market acts through data. Actively celebrating or guarding against neighborhood change in online discourses is a well-studied part of gentrification.[82] These discourses have real consequences: welcoming more wealthy residents into the neighborhood, putting pressure on low-income residents to somehow conform to a new normal or leave, and drawing attention from speculators, which can raise the home values in a neighborhood even further.

The discourse of gentrification introduces alternative ways of thinking about the potential impacts of changing home values in a neighborhood. Nevertheless, discourse alone does not make data actionable at a structural level. Policy depends on algorithm-like rules. For example, tax assessments are a result of algorithms: rule-based procedures that take existing data as inputs. Consider this description from the Georgia Department of Revenue of how property taxes are calculated:

> The assessed value—40 percent of the fair market value—of a house that is worth \$100,000 is \$40,000. In a county where the millage rate is 25 mills (1 mill = \$1,000 of assessed value) the property tax on that house would be \$1,000; \$25 for every \$1,000 of assessed value or \$25 multiplied by 40 is \$1,000.[83]

This is the language of procedure. It is one of many algorithms involved at the level of policy. At around \$300 for a custom "fee" appraisal, the county can't afford to independently assess the value of every home. It instead relies on comparable sales to calculate the "fair market value" of every home as well as the associated tax bill. These "mass appraisals" depend on techniques similar to those employed in Zillow's Zestimate.[84] As home valuations increase in a neighborhood, algorithms dictate that so do taxes, regardless of what the owner originally paid. For homeowners, rising home values often mean a larger tax burden. Thus without intending it, when my partner and I bought a home in a market with rising prices, we increased the cost of living for our neighbors.[85] If they can't pay their new bills, they may have to leave, willingly by selling their homes, or forcibly through eviction or foreclosure.

There are a few ways to reduce the property tax burden of individuals. These are algorithmically determined as well. Rules govern the conditions under which your taxes might be adjusted. For instance, if you live in your home (a rule-based determination), you can apply for a homestead tax exemption:

> The home of each resident of Georgia that is actually occupied and used as the primary residence by the owner may be granted a \$2,000 exemption from county and school taxes except for school taxes levied by municipalities and except to pay interest on and to retire bonded indebtedness. The \$2,000 is deducted from the 40% assessed value of the homestead.[86]

This only applies to a portion of a home's value. Beyond the initial exemption, homeowners are unprotected from tax increases. There are other exemptions based on age, but nothing at the time of this writing related to income, at least not in the local context of Georgia.[87] Finally, you can contest the county's appraisal of your home's value if you follow a set of predetermined procedures. But learning about and taking advantage of these tax relief measures takes time as well as resources that not everyone has.

All these regulating algorithms contribute to the complex civic context for thinking about and acting on housing data. Understanding the broader impact of these rules is not easy. Some argue that the homestead exemption act—originally instated to support homeownership—shifts the burden of taxes, necessary for schools and other local services, to those who can't afford to buy. Landlords, who aren't able to claim the exemptions themselves, routinely transfer the rising cost of taxes directly to their tenants.

In response to this regulatory setting, some organizers and researchers are fighting for more renters' rights.[88] Organizers at the Housing Justice League work with renters in Atlanta, mostly people of color, who have been priced out of the communities that they grew up in. They have recently completed a major public report on gentrification and its effects in Intown Atlanta.[89] May Helen Johnson, an elderly low-income resident profiled in the report, expresses concerns about two new large-scale developments near her home: the Mercedes-Benz Stadium, home to the Atlanta Falcons and United FC, and the public-private infrastructure project known as the Atlanta BeltLine.[90] What will they mean for her community?

> With the stadium on one side and the BeltLine on the other, it feels like we're being compressed between these two giants and my thing is what are they gonna bring to the neighborhood, what are they gonna offer us? Are we gonna be able to stay in the neighborhood or are we gonna be able to rent, to buy, to play, to stay, to worship in the neighborhood?

What would it take to enact regulatory policies in Atlanta that protect low-income residents from the inevitable outcomes of a market on the rise? "A crisis is hitting renters. We need data to declare a renter's state of emergency," the organizers explain.[91] Their policy suggestions include procedurally driven tax relief for developers who are willing to build and maintain low-income units.

The same organizers argue that a critical reading of existing data isn't enough to change the tide. They are amassing counterdata to fight gentrification. Organizers contend that "we need data on how many people are being displaced. We need data on their mental, emotional, and physical health. Who's being displaced and what is the consequence of that? We need data to show that there is mass displacement that is causing great suffering."[92] Together, organizers and residents from gentrifying neighborhoods across the city are working to fill in the missing context, which shows that the effects of the 2007 housing crisis have not abated.

As a contribution to the Housing Justice League report, I worked with students from Georgia Tech to create an online interactive map that displays this sort of counterdata (figure 5.7). The map visualizes demographic indicators of gentrification identified in the report—percent change in median income, college education, and "white share" of the population—for neighborhoods (defined by census tracts) along the current and proposed path of the Atlanta BeltLine, which is currently under construction along a loop of disused railroad tracks that circumvent the city, stitching together some of its most historic neighborhoods.[93] The last of these indicators suggests (inversely) how many people of color have been displaced from an area. The squares on the map represent census tracts along the BeltLine: red denotes an increase in the indicator selected, and blue shows a decrease. The squares in gray represent other tracts in the surrounding county. For individual neighborhoods, of which the Old Fourth Ward is but one of many, the map is a starting point for a different conversation about the future of Atlanta among Housing Justice League members and other marginalized subjects of housing data in the city.

Using visual, discursive, and algorithmic processes, housing data can be critically recontextualized to counter consumeristic desires endemic to the culture of real estate and instead promote a common good. In Atlanta, where low-cost housing is disappearing across Intown neighborhoods, there is an imperative to shift the current context in which we act on housing data. When placed in an operational context, data do not only represent an existing condition in housing; they enable practices that benefit some and harm others. But in order to counter the consumerist uses of data shaping Atlanta's future, organizers must learn how to effectively present data within a civic context, perhaps through interfaces that have yet to be invented.

Can interfaces cause friction in the market rather than kill it? Can they help us critically reflect on how the culture of real estate has evolved? Can they support us in talking more frankly about gentrification? Can they build support for another sort of algorithm: rule-based regulations with the power to restrain market fluctuations along with their effects on low-income homeowners and renters? These are questions that do not have easy answers. Yet I believe that for such interfaces to work—not in some generic sense, but in real places like Atlanta—they must confront the overarching claim of this book: all data are local.

CONCLUSION

In the rush to make publicly available data more accessible and actionable, companies within the interface economy seek to recontextualize it. As Zillow's interface demonstrates, frequently the context for data is not simply representational: an account of the setting in which the data were made. Nor is context interactional: the spontaneous result of an engagement with data. Rather, putting data in context is often operational: the result of connecting data to an existing knowledge system, defined by a combination of practices, processes, concepts, and affordances brought together in an interface meant to support the data's use. Through visual, discursive, and algorithmic devices, Zillow has constructed an attention-grabbing interface that supports the use of data to buy, rent, or sell housing. The operational context that Zillow has assembled, however, also reifies dominant and deeply problematic relationships inherent to our market-based culture of real estate.

While technologists and designers at Zillow are working to establish frictionless settings in which consumers might make the best personal choices based on housing data, housing organizers seek to construct another context for data—one that calls

A BeltLine for all?

This map visualizes demographic indicators of gentrification in neighborhoods (defined by census tracts) along the current and proposed path of the Atlanta Beltline, an “urban redevelopment” project under construction along a loop of disused railroad tracks that circumvent the city, stitching together some of its most historic neighborhoods.

show

dekalb county
fulton county
near beltline
highways
marta

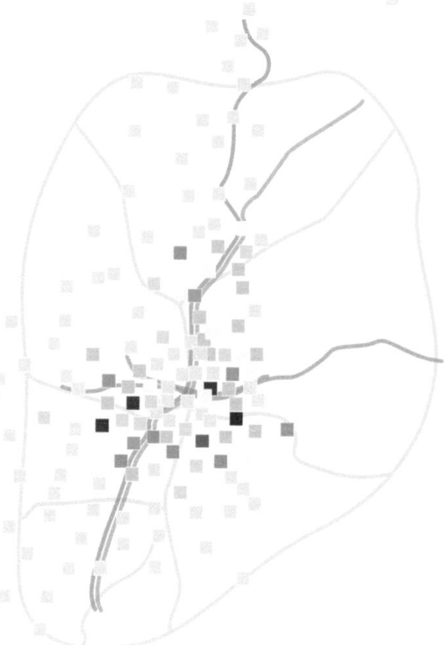

5.7

An online interface visualizing demographic indicators of gentrification (percent change in median income, college education, and “white share” of the population) in census tracts along the Atlanta BeltLine. All data are from the US Census Bureau’s American Community Survey. Image by the author, Christopher Polack, and Peter Polack.

sort 2010 – 2015

median income ◉ + 53%
college education ○ + 10%
white share ○ + 18%

Census Tract 50
Reynoldstown

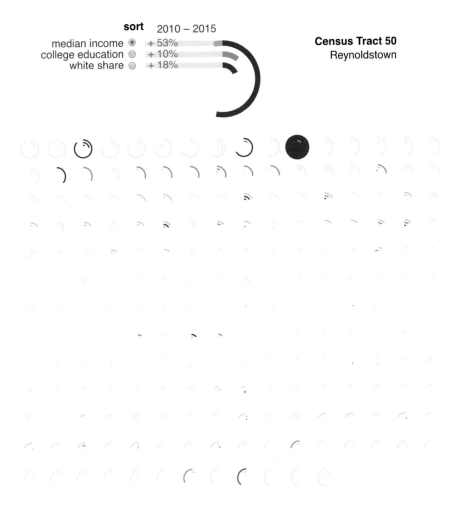

the market into question. These conflicting contexts—one consumerist, and the other civic—enable different ways of imagining and enacting the future of Atlanta through data. They both implicitly accept that data are now a necessary tool for addressing the problem of housing, which has reached a scale that would be difficult to contemplate otherwise. Indeed, the context of housing data has become a site of contestation, which will determine how the city evolves, and for whose benefit.

In this chapter my explanation of interfaces as recontextualizations of data is illustrated by cases from the domain of housing, where the stakes for accessibility, interpretation, and action are high, particularly in my own corner of the world. But there are many other domains from which important examples might be drawn: health, crime, and climate change, to name a few. In all of these areas, practitioners who design interfaces to data do not act autonomously. Rather, they must connect with cultures of use to support data-driven action. For data are made actionable in culturally rooted knowledge systems: consumerist, civic, or otherwise.

6

MODELS OF LOCAL PRACTICE

INDEXES TO LOCAL KNOWLEDGE

Throughout this book, in chapters on data's place attachments, infrastructural conflicts, algorithmic entanglements, and recontextualizing interfaces, I have sought to bring together the components of an overarching theoretical proposition expressed in the title, *All Data Are Local*. Many scholars have written about the locality of practices with scientific data over the past few decades.[1] It is past due time to test this general proposition with the everyday instances of data all around us: public records, the contents of cultural repositories, the results of commercial algorithms, and market-based values, all of which we too often mistakenly rely on as facts. As data become more ubiquitous, we must acknowledge that *all our amassed records are no more than indexes to local knowledge*. This point, first presented at the beginning of the book, is the summative principle of the chapter that follows.

What are the implications of this general proposition, and how might they manifest in practice? Moreover, what is to be gained by adopting a local sensibility? These questions should be of relevance to anyone struggling to find meaningful patterns in the maddening variety of currently available data. In order to present contemporary examples of local practice, this chapter takes on a slightly different style and scope. It draws together principles from throughout the book and synthesizes them into an equal number of lessons:

- All data are local
- Data have place attachments
- Data have heterogeneous sources
- Data and algorithms are entangled
- Interfaces recontextualize data
- Data are indexes to local knowledge

- Look at the data setting, not just the data set
- Make place part of data presentation
- Take a comparative approach to data analysis
- Challenge normative algorithms using counterdata
- Create interfaces that cause friction
- Use data to build relationships

In what follows, these lessons are further illustrated by intentionally unconventional models of practice, chosen specifically to overturn assumptions about data's origins and their uses. These include a data documentary (*Bear 71*), data performance (*A Sort of Joy*), data correspondence (*Dear Data*), counteralgorithm initiative (Grassroots Mapping), data narration ("Slave Revolt in Jamaica, 1760–1761: A Cartographic Narrative"), and data-driven movement (Anti-Eviction Mapping Project). They are intentionally different from the examples of visualization that I have prepared for the pages of this book, demonstrating modes of engagement with data not limited by the parameters of the printed page. Although these projects have all been published before, my goal here is to reframe how they are conventionally understood in order to further elucidate what local can mean in relationship to data.

Beyond introducing these specific examples, however, this chapter makes a claim approached only implicitly in the rest of the book: that practices of design can enable critical reflection on the local conditions of data. In these projects and throughout the book, I take design to mean—in the tradition of economist turned design theorist Herbert Simon—a transformation of data from an "existing situation" to a "preferred one."[2] Yet design can also be something responsive to local conditions. As philosopher Donald Schön explains, it can be a "conversation with the materials of the situation."[3] In the examples that I have chosen, the materials are not only data but a more encompassing data setting too. The conversations, meanwhile, are variously programmed, spoken, drawn, mapped, and performed. This definition of design might be criticized for its overly general character. In fact, some of the examples included below might come across to the reader as art or activism. Regardless of what we call them, they fit well within the scope of creative practices by which I believe all readers can and should be informed.

By *critical reflection*, a phrase first put forward in the introduction, I mean a process by which the interwoven social and technical dynamics of data are made visible and accessible to judgment.[4] Critical reflection allows us to ask, What are the otherwise-invisible attachments, values, absences, and biases in data? Without the capacity for critical reflection on data, we risk adopting the unspoken assumptions that are embedded in all modes of data creation, processing, and use.

Many authors consider the ways in which design and other allied forms of making can aid in critical reflection. A pioneer of "critical technical practices," Phil Agre develops exercises intended to discover inherent limitations in computer systems.[5] Instead of seeing a technical limitation as a problem to be solved, he suggests that we should learn to diagnose it on a deeper level. For such limitations may be evidence of systemic misconceptions that need to be addressed on a social or philosophical level.

Comparable work in ensuing years, strongly influenced by the public-facing "critical designs" of Anthony Dunne and Fiona Raby, has become more accessible.[6] Critical design exhibitions and interventions present "contestational" objects, which "challenge and offer alternatives" to the technological norms of the moment.[7] In this way, design can be a rhetorical strategy in addition to a means of inquiry. Moreover, design can transcend logocentric approaches, in which words alone are expected to gather evidence and support claims.[8]

Below I share models of critical reflection that move beyond the sketches and diagrams that I have devised for this book project. The examples provided in the previous chapters only gesture at the full possibilities of design as a critical mode for inquiry and claims making that can address the local conditions of data. Although the history of design with data is long, I will focus on illustrations that are contemporary, produced just before or during the writing of this book. As these instances attest, I am not the first designer to attend to the local characteristics of data. But I hope to finally make evident a local sensibility toward data that has, until now, only been addressed piecemeal or understood intuitively by those who work closely as well as creatively with data.

PRINCIPLES TO PRACTICES

Many essays and books struggle to connect theoretical principles to practical ends in their last few pages.[9] Readers invested in the future of information technologies may expect to see solutions fill this space: lessons on how to fix a problem introduced in the main body of the text or predictions for the next generation of incremental progress. Such prescriptions are challenging to produce and risk coming across as a form of "discount theory": easily accessible but of little lasting value.[10] Instead of offering simplistic implications for practice, this chapter presents a set of concrete models that manifest the six overarching principles of the book without closing down a range of possible interpretations and adaptations.[11]

Such models are used widely in design education to convey implicit lessons.[12] Contrary to some of their counterparts in departments of engineering, design students are not taught a strict set of rules or procedures. Rather, they are empowered by models that show, versus simply tell, what good design means. These models are not meant to be replicated for they have a local specificity that must be entirely rethought for new circumstances. Too often, data are displayed using design templates, which are falsely presented as universal perspectives on data that we now know are local.[13] In contrast, the six models below take little for granted.

To understand how models can carry theory, consider the "Norman door." Years ago, a student printed those words on a small label and posted it next to a set of glass doors that mark the entrance to Georgia Tech's Program in Digital Media, where I teach. A classic example from Donald Norman's 1988 book *The Design of Everyday Things*, a Norman door is any ambiguous egress that is missing appropriate labeling for "push" or "pull," thereby baffling its users. Our local version of the Norman door is a hinged glass door with identical handles on either face. To the evident frustration of every visitor who mistakenly pushes when they should pull or vice versa, the door only opens in one direction.

This egress demonstrates the validity of one of Norman's core principles of human-centered design: "make things visible." But people are more likely to remember the concrete example of the door, originally written as an account of Norman's friend, who found himself trapped between two sets of such doors without being able to see his way out.

> A swinging door has two sides. One contains the supporting pillar and the hinge, the other is unsupported. To open the door, you must push on the unsupported edge. If you push on the hinge side, nothing happens. In this case, the designer aimed for beauty, not utility. No distracting lines, no visible pillars, no visible hinges. So how can the ordinary user know which side to push on?[14]

The Norman door has become a proxy for a principle. Similarly, Norman's other principles, such as "get the mappings right," "exploit the power of constraints," and "design for error," are not as memorable as the examples themselves. Many designers cannot look at Legos, for instance, without thinking about Norman's lesson on constraints. Legos are satisfying in part because constraints in their form—the tiny pegs—make it clear how they need to fit together. As I shift from empirically and theoretically oriented chapters on the local conditions of various data sets, toward more practical lessons, I am reminded of Norman's models, and want to leave the reader with compelling and transferable manifestations of the ideas in this book.

Six local models of practice are explained below. Each model is described as an expression of one of the principles introduced earlier in the book. Alongside, I offer some advice about how to adapt each model for other local conditions.

LOOK AT THE DATA SETTING, NOT JUST THE DATA SET

Data are created by humans and their dutiful machines in a place, at a time, often from within an organization, using the tools at hand, for audiences that are believed to matter. Indeed, data are grounded in varied data settings that are not accounted for in their presentation as data sets. Data sets instead are taken to be closed, self-contained, independently interpretable collections that can be transferred anywhere, copied anytime, and decoded by almost anyone. Throughout the book, my examples from the Arnold Arboretum, DPLA, NewsScape, and Zillow serve primarily to reinforce this point. As I hope the reader will appreciate by now, data cannot speak for themselves.

Turning to preexisting design work, one project that I see as a strong model for the exploration of data settings is *Bear 71*, a twenty-minute documentary about the eponymously numbered grizzly bear (figure 6.1). The project uses data (over one million photos) from creature cams: cameras that operate autonomously in order to capture rough images of animals in their wild environs. These instruments are triggered by movement and, thus, can function at times and places where humans cannot. *Bear 71* was developed by Jeremy Mendes and Leanne Allison to allow audiences to interactively follow the story of a single animal across a virtual 3-D map of Canada's Banff National Park.

Moreover, the project is meant to raise awareness about "intersection(s) of animals, humans and technology."[15] In doing so, it challenges our assumption that data are a product of human invention alone.

Bear 71 began with a collection of photographs taken by the aforementioned cameras stationed throughout Banff. Its creators knew that the low resolution and grainy quality of these photos made them inappropriate for a traditional big screen documentary experience. Instead, they crafted one for smaller, more intimate screens. By acknowledging and foregrounding the inherent qualities of their photographic data, the creators of *Bear 71* bring audiences into a data setting. This setting is represented as a stylized topography that can be interactively explored, not from a god's-eye perspective but rather on a ground in which "the boundaries of the wired world are blurred."[16] After donning virtual reality goggles, audiences occupy a place in the virtual landscape from which they can interact with a network of fifteen cameras as well as data on animals tracked by radio transmission collars, such as Bear 71 herself, and other phenomena of human creation, such as (spoiler) the train that ends her life. The film addresses larger themes about how surveillance is collected and experienced by making that experience strange through the eyes, ears, and voice of Bear 71.

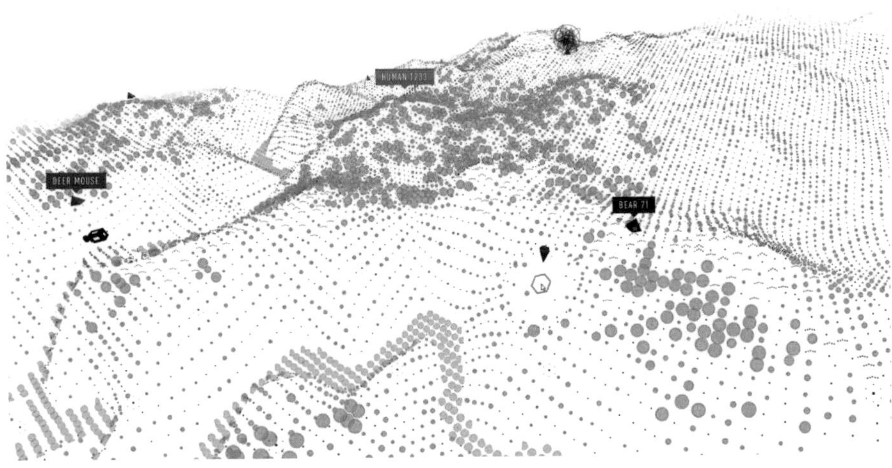

6.1
The data documentary, *Bear 71*. Courtesy of the authors, Jeremy Mendes and Leanne Allison.

I call this piece a data documentary because not only does it use data in the course of storytelling but rather data—in this case, an overwhelming collection of wild cam pics—are among the primary subjects of the story.[17] *Bear 71* helps us see these data within the local conditions in which they were created and reflect on the role of surveillance data more generally on a planet so littered with sensors that Jennifer Gabrys has dubbed it "Program Earth."[18]

Adapting this model for other data means looking at the social and technical setting in which the data are made and managed. In *Bear 71*, the data setting is the national park where those grainy pictures originated. By representing both the pictures and landscape in digital form, the story becomes about something more encompassing than simply data. *Bear 71* is about evolving relationships among the technologies of data collection, the human society that created those technologies, and the nonhuman denizens that are subject to them. This approach helps us think about data as part of a larger system that needs to be analyzed holistically. Representing data settings requires a multipronged approach, bringing together descriptions of a wide variety of actors.

In order to conceptualize a data setting, it might help to consider some of the following questions. Where are the data collected? Are there multiple sites? If so, how are they different? How might we understand the organization of the collection site (or sites) in ways that would enhance our appreciation of their data? What instruments are used to collect the data, and what kinds of limitations do they have? Who is managing the instruments that collect the data? What aspects of their experience are captured or obscured by the instruments? Who are the subjects of data collection, such as Bear 71 herself? What other actors are accidentally captured in the data? In *Bear 71*, the title character is one of many different animals that appear periodically on the creature cams. Finally, which actors are not directly captured in the data but nevertheless are important for understanding the setting, such as the train that ended Bear 71's life?

MAKE PLACE PART OF DATA PRESENTATION

Place is a common field in many data sets. It is the site in which we encounter data and an important part of the settings in which data are made. Furthermore, places are increasingly understood in terms of the data available about them. As a dimension of data, place is typically reduced to a geolocation. In chapter 1, I took the reader on a tour of the Arnold Arboretum to illuminate the many ways in which data hold other kinds of attachments to place. Place surrounds and pervades data in ways that may seem obvious, but that have crucial nuances to which we must be sensitive. Despite the potential richness of thinking about the place of data, we frequently treat data as virtual and ubiquitous, even fearing that data can distract us from attending to the place we are in. As a model of a place-based engagement with data, consider the unusual data performance titled *A Sort of Joy: Thousands of Exhausted Things* (figure 6.2).

It's 1:32pm. A woman in a black dress leans against the edge of a doorway between rooms in MoMA's second floor galleries. Swatches of rotating light and the ting-tang of a Gamelan orchestra from the installation behind her bleed past her, out into the room that she's facing.

"Fuck Off," she mutters.

A few faces in the crowd turn towards her, but most either didn't hear, or pretended that they didn't hear. The woman continues, undeterred.

"Where's My Fucking Peanut?"

"Shut The Fuck Up."

"I Shit Crystals for you, David."

Despite this impressive string of obscenity, the gallery goers' attention is mostly directed towards the middle of the room, where a group of five people have just burst into song.

Over the next forty minutes, this group of six performers will speak (and sing) in a strange language—every word they say will be taken verbatim from the collections database. And yet it will not come off as if they are listing a litany of titles; instead they will engage in complex patterns of call & response, performing a combination of carefully choreographed exchanges and loosely-defined scenes, often balanced at the edge of chaos and absurdity.[19]

6.2

The data performance, *A Sort of Joy: Thousands of Exhausted Things*. Courtesy of the authors, Office for Creative Research.

This text was written by artist and designer Jer Thorp, one of the authors of the performance. The event was organized in 2015 by Thorp's former studio, the Office of Creative Research, in collaboration with the theater group Elevator Repair Service as part of a residency at the Museum of Modern Art (MoMA) in New York City. In contrast to conventional forms of visualization applied to data, *A Sort of Joy* is aural and ephemeral. It consists of a series of readings from "sorted" lists of data about art objects held by the museum. The lists, generated by the Office of Creative Research from MoMA's collections data, include the names of artists and titles of artworks (as illustrated above), among other fields listed in the museum's records. The performance does not reveal how these data are stored or used by museum staff. Instead, the data are taken as a starting point for a creative act—one that conjures titles, names, dates, and other descriptors in a provocative place: a MoMA gallery in front of various works of art, most not directly related to the data being read.

A Sort of Joy is a local practice that uses the data to bring an otherwise-invisible network of associations to light in a public place, but not without adding a layer of expressiveness through the choices, voices, and bodily gestures of the performers. Some of these readings are funny, as in the initial example. Others are poetic or critical.

For instance, a list of first names is read. They belong to artists with work in MoMA's collection and are sorted by frequency. The top thirty are as follows: "John, Robert, David, Paul, Richard, William, Peter, Charles, Michael, James, George, Jean, Hans, Thomas, Walter, Edward, Jan, Joseph, Martin, Mark, José, Louis, Frank, Otto, Max, Stephen, Jack, Henry, Henri, and Alfred."[20]

Hearing this reading of artists' names, one can't help but be struck by the absence of women. Who can be an artist? What is art? What is a museum? These are all questions that might arise in the minds of audiences as *A Sort of Joy* unfolds. The project challenges us to see the museum as data and, in doing so, to see the place of data presentation as a subject of attention and inquiry.

There are many ways in which place is enrolled in *A Sort of Joy*: as the site of data's production and engagement as well as a subject that can be portrayed through data. This project does not engage with the "place values" in data (i.e., where these works come from) or data that are out of place (those errors and anomalies in the data that might complicate our simple understanding of their origins). Such projects need not address every dimension of place. It calls attention to the place of data in a productive way because it makes us think about where we are encountering MoMA's collections data as well as the kind of place the museum is.

Adapting this model involves considering the places where we want data to be encountered. In order to do so, we might try answering some basic questions. What locations and times of presentation would resonate or produce friction with the data, as in the case of *A Sort of Joy*? Can one of these places be reimagined as a site of intervention? How do dimensions such as geometry, orientation, or adjacency shape the site in notable ways? What conditions or cycles is the site subject to? Does the site include people, animals, or other kinds of social activity? If so, what patterns of behavior might characterize it? How can we imagine relating the data and site: through superimposition, juxtaposition, or some other technique? Is there a convergence between the data and the site, or do we seek to produce some form of tension? How might the data be manifest in ways that make sense for the place: as projections, sounds, objects, or in some other form?

TAKE A COMPARATIVE APPROACH TO DATA ANALYSIS

Data sets of high volume and variety are often composite collections created across disparate times and places. In fact, we would better understand all data by thinking about them as composites. Furthermore, we should explore how data are created otherwise, in settings with different guiding values and norms. The local dimensions of data are most evident when different sources are brought into dialogue. In chapter 2, I explained how the DPLA provides an opportunity to see data from a variety of institutional origins. My visualizations of the DPLA, focused initially on data schemata and then data structures, illuminated the varied ways in which libraries, museums, and archives, among other sources, make data their own, in their own ways.

This comparative sensibility is at the center of a project by information designers Giorgia Lupi and Stefanie Posavec, titled *Dear Data*, which I first learned about during a visit to the Eyeo Festival in Minneapolis (figures 6.3 and 6.4). The two information designers met at the same festival several years before launching the project, but they lived on opposite sides of the Atlantic: Lupi is in New York City, and Posavec is in London.

In order to get to know one another, they used an old-fashioned form of correspondence, yet with a contemporary twist: they sent one another postcards with hand-drawn visualizations based on data that they collected about their lives. "Every Monday we chose a particular subject on which to collect data about ourselves for the whole week: how often we complained, or the times when we felt envious; when we came into physical contact and with whom; the sounds we heard around us."[21]

At the end of each week, they converted their records into tiny but evocative hand-drawn visualizations, filling only a single side of a four- by six-inch postcard. The other side was reserved for the shipping address, a legend for the visualization, and occasionally a short apology regarding something that went wrong during the process of data collection. Over the course of the project, Lupi and Posavec slowly opened their lives to one another (and later to the rest of us) through the lens of seemingly mundane everyday events, but also through their own expressive choices about how to capture and convey these minutiae. In a book about the experience, also called *Dear Data*, they write, "We are all creating data just by living."[22] But data aren't simply a by-product of life. They are deliberately designed, as much as any visualization, in order to represent selected events and experiences.

6.3

The data correspondence, *Dear Data*. Courtesy of the author, Giorgia Lupi.

"DEAR DATA
WEEK 01: WHAT'S THE TIME?

HOW TO READ IT:

- each symbol represents every moment I glanced at the clock, grouped by hours of the day.
- Different symbols and attributes represent WHY and HOW I checked the time.

1AM
12 AM
M T W T F S S
- days of the week -

SYMBOLS

- o on purpose: wanted to know what the time was.
- • Just glanced: on a phone, mac or else
- ⊠ Because I thought of this project.
- x I thought "Don't look!" But I did.
- I Because I was Bored
- Ir Because I was Hungry
- Δ Heard somebody saying the time aloud.

ATTRIBUTES

- f**ck! I'm late!
- oh, ok. I'm fine.
- analog support (i.e. wrist watch)
- alarm clock rang.
- glanced at the clock while texting or emailing with Stefanie. ☺

SEND TO:
STEFANIE POSAVEC
LONDON
[UK]
ENGLAND

00122$0001

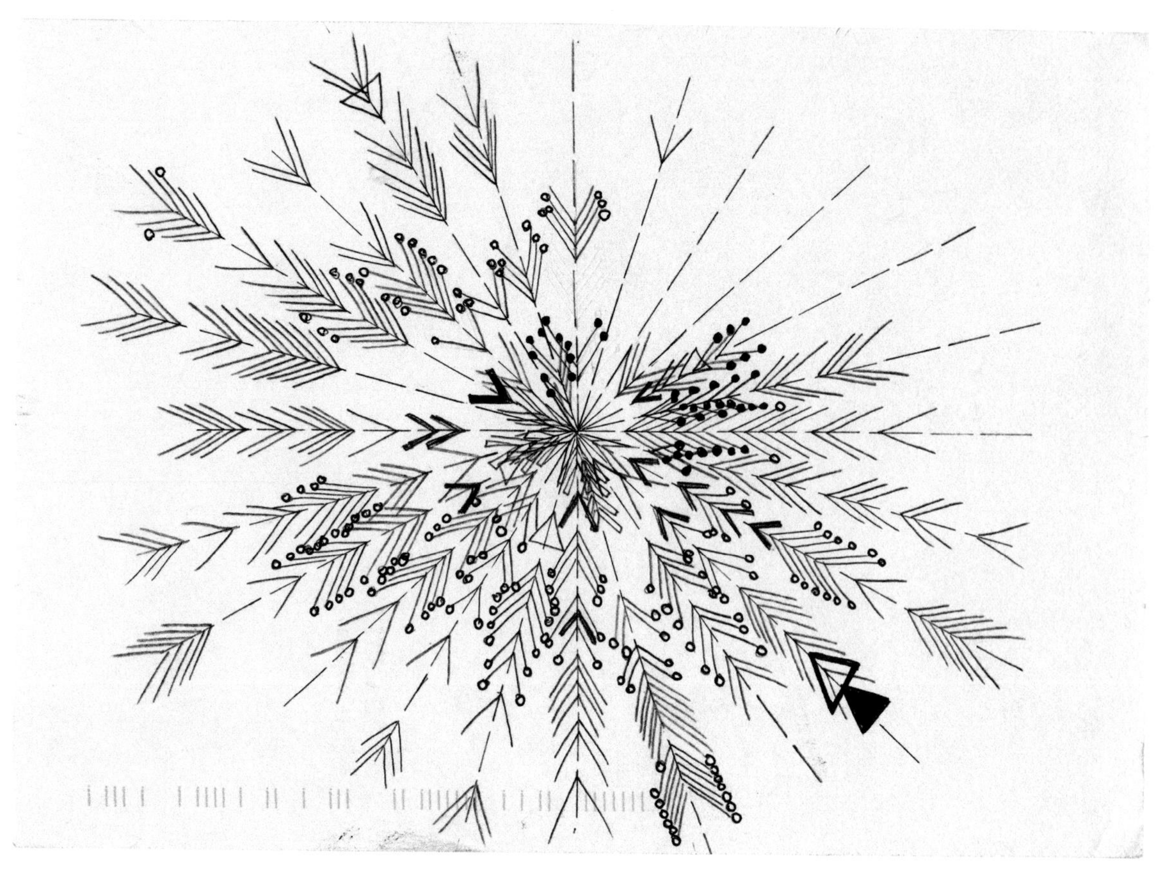

6.4

The data correspondence, *Dear Data*.
Courtesy of the author, Stefanie Posavec.

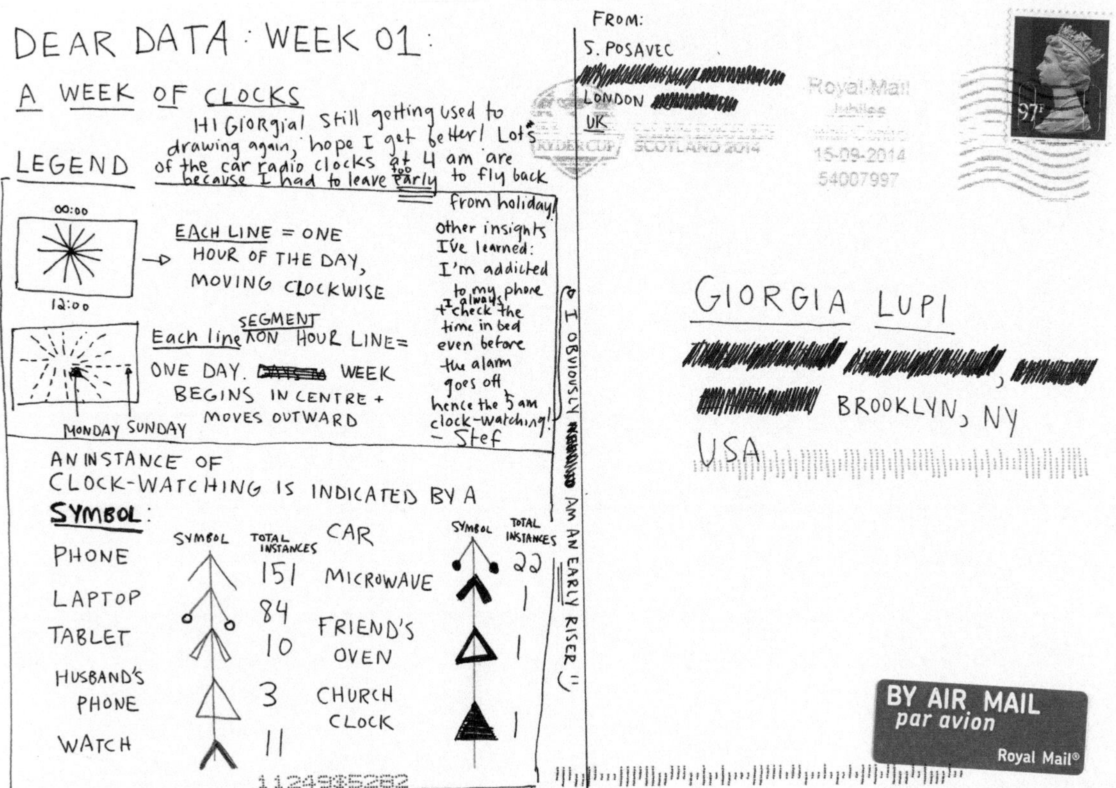

DEAR DATA : WEEK 01 :
A WEEK OF CLOCKS
HI Giorgia! Still getting used to drawing again, hope I get better! Lots of the car radio clocks at 4 am are because I had to leave too early to fly back from holiday!
LEGEND
00:00
12:00
EACH LINE = ONE HOUR OF THE DAY, MOVING CLOCKWISE
Each line SEGMENT ON HOUR LINE = ONE DAY. WEEK BEGINS IN CENTRE + MOVES OUTWARD
MONDAY SUNDAY
Other insights I've learned: I'm addicted to my phone + I always check the time in bed even before the alarm goes off hence the 5 am clock-watching! - Stef
I OBVIOUSLY AM AN EARLY RISER :)
AN INSTANCE OF CLOCK-WATCHING IS INDICATED BY A SYMBOL:
SYMBOL
TOTAL INSTANCES
PHONE 151
LAPTOP 84
TABLET 10
HUSBAND'S PHONE 3
WATCH 11
CAR 22
MICROWAVE 1
FRIEND'S OVEN 1
CHURCH CLOCK 1
FROM:
S. POSAVEC
LONDON
UK
Royal Mail
15-09-2014
SCOTLAND 2014
RYDER CUP
97p
GIORGIA LUPI
BROOKLYN, NY
USA
BY AIR MAIL
par avion
Royal Mail®

I do not know if the creators anticipated the degree to which their approaches to data collection each week would differ and that those choices would be as expressive as any other characteristic of their work, particularly when looked at in comparison. Still, the comparative nature of the project gives it a strength and personality that other well-known examples of artists' and designers' self-reporting initiatives don't have.[23]

For instance, their first theme was "a week of clocks," reminiscent of the example in chapter 2 about the representation of time in library records. In *Dear Data*, however, the method of recording is not a function of institutional bureaucracy but rather one of individual experience and, more pointedly, anxiety about the passage of time. Lupi noted every instance in which she checked the time and annotated it with what she was feeling: "I am so late!" or "I am bored" or just "super hungry!"[24] We don't know what her original records look like, but her data visualization reveals time checks organized by hour and day, lined up one after another to present the week as evidence of her attention and attitude toward time. Meanwhile, Posavec's attention was focused on where she checked the time: on her laptop (84 times), her watch (11 times), a friend's oven (1 time), and her phone (151 times), leading her to the exasperating conclusion that she is "addicted to her phone."[25]

Their visual representations differ as well. Lupi's glances at the time are represented as what seem like little ticks against her, adding up over the course of the week to a clinical tally of microconfessions. Posavec's reports are organized radially, as if on the face of a clock, with V-shaped notches (perhaps viewing cones) directed out from the center on each hour. Neither is particularly precise or legible; Lupi's rows of unannotated hours give no indication of when they begin or end, and we don't know if Posavec's times are a.m. or p.m. These are clearly meant to be expressive rather than standardized—as most clocking in usually is. "'We've always conceived Dear Data as a "personal documentary" rather than a quantified-self project which is a subtle—but important—distinction,' explains Lupi."[26] Indeed, their choices about how to record time data and how to codify it in these visualizations say more than a formal reading of the visualizations can reveal.

Dear Data demonstrates how informative it is to see data collected and represented differently. Those differences are important expressive choices and, in this case, reflections of personal identity. More clearly than any recent data project that I have seen, the work of these two designers reveals the idiosyncrasies of data. Their lessons about what they call "small data" go for all data. Data are always agglomerations of local tallies.

Adapting this model requires engaging at least two data settings; this is double the work of a traditional data visualization project. By embarking on such a task, we are challenging ourselves to discover the following. Where would we find examples of similar data from different places, times, or perspectives? What are the commonalities and differences in these sources? How might they be juxtaposed or used to complement one another? What do they reveal about how data collection can be expressive in its own right? What do they suggest about the identities of their collectors (positive or negative)? What is at stake in privileging one data set over the other? Do two data sets raise or lower levels of uncertainty?

CHALLENGE NORMATIVE ALGORITHMS BY CREATING COUNTERDATA

Data and algorithms are too often thought of as discrete elements of computing, with parallel but independent histories. In chapter 4, I illustrated their interdependence using the examples of news data and NLP algorithms. The two are materially and historically entangled in subtle, largely invisible ways. As we confront new data sets, we should be mindful of the algorithms involved in creating and/or processing those data. The reverse also holds true. We must not take the universality of algorithms for granted. It aids in our understanding of an algorithm when we consider the data it was tested with or trained on. Extending from this insight, researchers and practitioners who wish to challenge the normative assumptions in existing algorithms must create their own counterdata—a contestational approach to data collection that was explained more fully in chapter 3.

One model of counterdata in practice is Grassroots Mapping (figure 6.5) by the Public Lab (first introduced in chapter 2). The project is a low-tech and local alternative to Google Earth, a widely used but little understood collection of heterogeneous image data from satellites and aerial photography mapped onto the geometry of the planet.[27] Grassroots Mapping, characterized by Public Lab as a form of "civic science," does not just produce a new data set. It offers a set of instructions for contesting existing data, necessitating only a digital camera, string, and helium balloon.[28] It is a recipe that my colleagues and I used to document Bussey Brook Meadow, our experience of which was also described in chapter 2. In comparison to Google Earth, the Public Lab's instruments are capable of producing aerial images that are higher resolution, more precisely targeted, and (because it is such an accessible method) under the control of the communities that have made them.

Grassroots Mapping has been framed as "counter" not only to conventional mapping technologies but hegemonic institutions too, such as the World Bank. Such institutions depend on geodata to perform probabilistic risk assessment. Seen internally as a pragmatic means of managing change, the World Bank's methods are contested by critics as top-down modes of administration that replicate historical arrangements of representation and control exerted by colonizers over colonial territories.

> In our projects to date, our tools have been used to contest official maps or rhetoric by enabling communities to map sites that are not included in official maps. In Lima, Peru, members of an informal settlement developed maps of their community as evidence of their habitation, while on the Gulf Coast of the [United States], locally produced maps of oil are being used to document damage that is underreported by the state.[29]

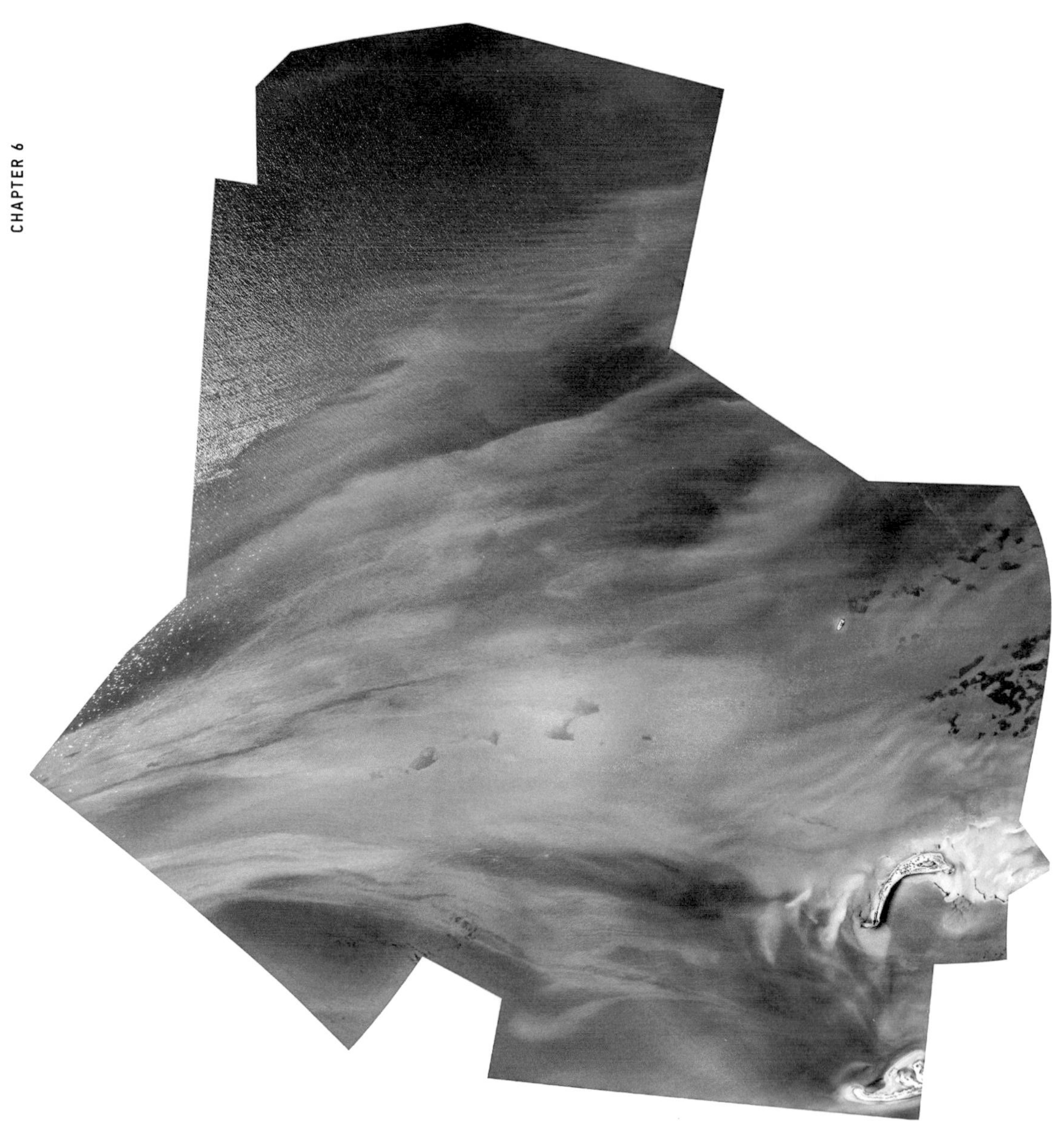

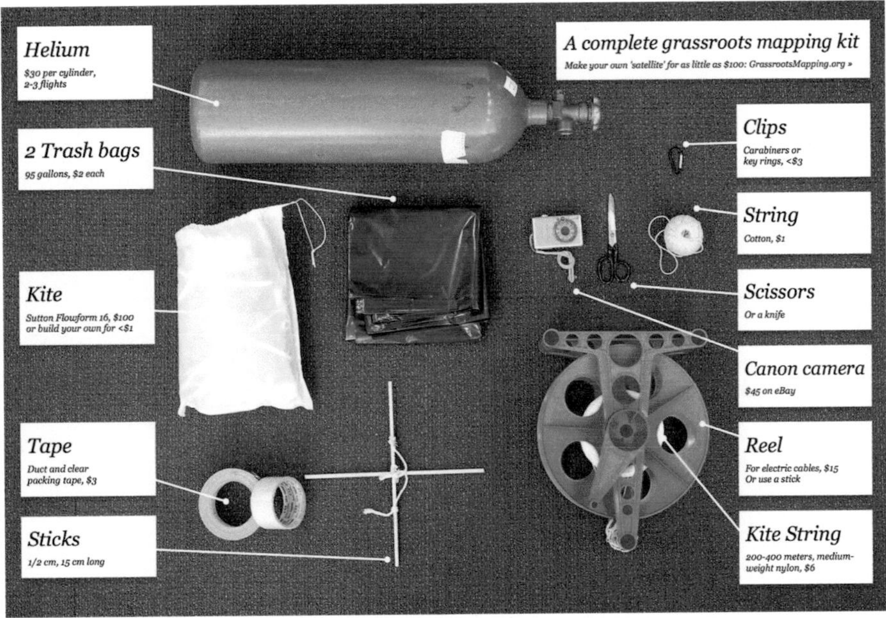

6.5

Parts from the counterdata initiative Grassroots Mapping. Photos were taken with a balloon and camera rig over the Gulf of Mexico after the BP oil spill. Courtesy of the authors, Public Lab contributors CC-BY-SA.

Grassroots Mapping not only produces better images for local communities. It calls attention to the limitations of all aerial image data, addresses privacy issues, and challenges the unequitable power relations that are a structural part of centralized, large-scale geodata collection. More recently, it offers an opportunity to challenge normative algorithms increasingly used in risk assessment.[30] Scholars in algorithm studies have already shown that current risk assessment algorithms are skewed by the kinds of data that they are trained on.[31] Creating new sources of data, as Grassroots Mapping does, can reveal the fragility and associated biases of existing algorithms, or even support the development of counteralgorithms.[32]

Beyond its activist uses, Grassroots Mapping can be utilized as an exercise in "situated learning," which challenges students to think critically about the political and social dimensions of any image capture process. Typically, courses in geographic information systems take aerial image data as "given," without considering their origins in military-industrial processing, centered on, say, mineral and petroleum discovery. Grassroots Mapping can be an entry point for teaching students about both the origins of aerial image data and how algorithms develop biases.

In many ways, Grassroots Mapping illustrates how challenging existing arrangements of data and algorithms can be a political act. Adapting this model requires an evaluation of the expectations of existing algorithms. What kind of data are they trained on? Creating counterdata or counteralgorithms requires asking the following questions. What networks of people, places, technologies, and funders support conventional data collection? What kind of power relations do they enact? Who and/or what is marginalized by these practices? Consider how data might be collected otherwise, by or about people and things that are otherwise left out of current collections. We must also think about how new data and algorithms might be presented in relationship to existing practices. Are they meant to fill in gaps in existing data, replace those data, or stand in juxtaposition? Finally, we must push beyond the typical notion of audience. Who might be able to replicate or take over the data collection process or the maintenance of algorithms as a means of self-expression?

CREATE INTERFACES THAT CAUSE FRICTION

Data are charged with meaning, significance, and opportunity when presented in context, by interfaces rooted in culturally based systems of knowledge. In chapter 5, I critically examined how Zillow's interface recontextualizes existing data collected from local settings across the United States. The resulting visual, discursive, and algorithmic context is one in which housing data are meant to be not only accessible but also actionable, although primarily for consumerist ends.

Alternatively, there are many examples of critical efforts to recontextualize data from distant places and times. "Slave Revolt in Jamaica, 1760–1761: A Cartographic

Narrative," for instance, relies on historical accounts to portray an insurrection involving more than one thousand black men and women enslaved within the eighteenth-century British Empire (figure 6.6). For this project, the historian Vincent Brown, and others, drew evidence from diaries, personal correspondence, newspapers, and administrative records created by white slavers—sources that by their very nature are skewed toward the slave owners' "insights, fears, hopes and desires."[33] Then the authors of the project recontextualized these sources to produce a counternarrative: a data-driven story about the greatest revolutionary effort in the history of the West Indies, a group of islands in the Caribbean originally colonized by Europeans in the late fifteenth century.

Over the course of eighteen months, the violent and now-shrouded events examined by "Slave Revolt" consumed much of Jamaica, beginning in St. Mary. Although the rebels managed to kill sixty slavers, almost ten times as many of their own died in the process. Another five hundred were dispersed from the island after the revolt was ended by the British counterinsurgency. Brown's project reads between the lines of the existing documentation to answer questions about what the revolt might have looked like from the perspective of the insurgents. What strategies did they employ? How did they coordinate? What might they have hoped to achieve?

"Slave Revolt" is presented as a historical narrative layered on eighteenth-century maps, originally drawn for Henry Moore, the lieutenant governor of Jamaica in 1756–1761. But instead of accepting these maps as given, which would lend explanatory power to colonial conceptions of the island, the project treats the maps as enemy terrain, which the rebellion made strategic use of. Topographic features, such as hills, forests, rivers, and estates, are georeferenced. This enables marches and massacres alike to be positioned in terms of approximate latitudes and longitudes. All the assembled evidence of events, stored in a database of locations, are animated as graphical lines of movement and estimated points of action to convey the political choices of the rebels, not spelled out directly by the original sources.

Unlike the other examples in this book, the project does not depend on a massive collection of data.[34] It is nevertheless a useful illustration because of the way in which it deals with the obvious attachments and uncertainties in its sources. For instance, Brown makes use of this excerpt from correspondence by a white "gentleman" to ascertain the vicinity and effects of the rebellion's early progress:

> At Day Light they appeared at Ballard Valley, towards whom Mr. Bayly approached, intending to expostulate with them, but firing five Shot at him he retreated, and both he and Mr. Cruikshank narrowly escaped with their Lives; they then fell upon the Overseer, Mr. McPherson, whom they killed, with three other white Men—Letter from a Gentleman at St. Mary, 14 April 1760.[35]

V.
SOULEVEMENT DES NEGRES
à la Jamaïque.
en 1759.
Dessiné par le Jeune.
Tom. III.
Gravé par David.

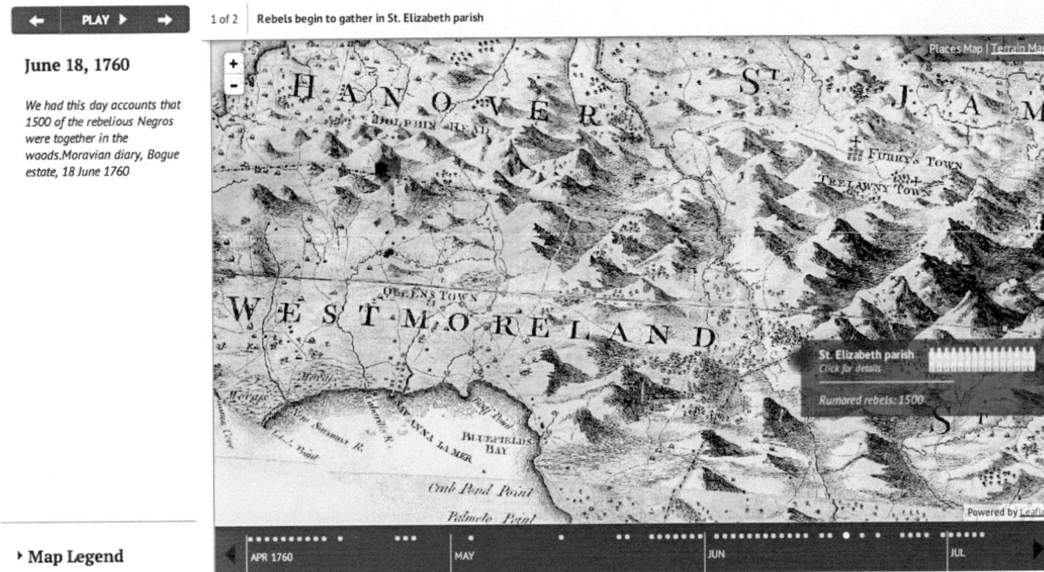

6.6

Visual elements of the data narrative "Slave Revolt in Jamaica, 1760–1761: A Cartographic Narrative." Courtesy of the author, Vincent Brown.

The local contingencies of the project's data present a pedagogical opportunity, argues Brown. Records of racist slaveholders and those in their service are not easily confused for the events that they describe. As such, their handling in "Slave Revolt" can help us consider how to grapple, more generally, with data that arise from inexorably local conditions.

In order to recontextualize and make productive use of what Brown calls a "debased database" of racist history, the project offers a critical interface built on many of the conventions explored in chapter 5. It creates a *visual* setting for data meant to reveal implicit choices that were not explicitly or reliably indicated in the original sources. It establishes a *discourse* that, instead of accepting sources based on the writings of slavers, helps us question them. Finally, it employs an algorithm to generate what Lev Manovich would call a data-driven narrative—"a cause-and-effect trajectory of seemingly unordered items (events)"—from the underlying database.[36]

Brown's narrative is one of many possible trajectories through the database that is meant to illuminate the logic of the revolt and open new questions about the ambitions of its leaders. Did they expect to prevail? Was the long-term plan to create their own settlements? Would they have appropriated the existing plantations? Could they take over the trade routes of their captors?

The project amplifies the lesson of chapter 5: an interface can introduce friction, challenging us to reconsider the meaning and interpretation of existing sources. Yet as a historical project, it is closer in its sources to the contents of the DPLA profiled in chapter 3. Brown struggles with many of the same challenges originally presented using examples from collections of cultural history, including historical assumptions, material constraints, and discriminatory categories. He writes eloquently about the difficulty of working with complex historical data:

> I share the tragic compulsion common to historians of the oppressed: though our accounts of slavery are distorted by the mediation of the sources, we persist in trying to explore and explain its past. Knowing that the truth is a receding horizon, we still set out to close the distance.[37]

Learning from this model involves, first, treating data sources as cultural artifacts, tainted by their own historical and material contingencies. Next we must find creative ways to recuperate and recontextualize them for the purposes of critical reflection. Refusing to take data as given—and thereby challenging the biases and values of their creators—can lead us to consider how interfaces might introduce friction in the presentation of data. Audiences of *frictional* interfaces should be prompted to ask questions. Whose data are these? Who do they speak for? What worldview do they embody? Whose voices are left out? Furthermore, such interfaces can actively counter damaging interpretations of data, and instead introduce visual, discursive, or algorithmic strategies that help us read old data sources in new and critical ways.

USE DATA TO BUILD RELATIONSHIPS

One of the reoccurring messages of this book is that data cannot stand alone as independent arbiters of important societal questions and controversies. Rather, data should be seen as a means with which to connect to and learn from those with actual local knowledge. Data can help draw knowledge together, but never replace it altogether. This principle is explored throughout the cases, in conversations with botanists at the Arnold Arboretum, librarians from contributing institutions to the DPLA, data journalists who grapple with new conceptions of the news, and organizers who seek to redefine the context in which we understand housing data.

All the aforementioned practices are of limited relevance if they do not help to reshape the social relationships around data. Engaging with data cannot simply be a matter of extracting information. Instead, data are an opportunity to connect to the people who make them, are subject to them, or through their own experience, know how to make use of them. Projects focused on the local aspects of data put people or other important actors like Bear 71 at the center.

One of the most successful instances of this approach to data is the Anti-Eviction Mapping Project—a project that became a community (figure 6.7). The Anti-Eviction Mapping Project is described by its members as "a data-visualization, data analysis, and storytelling collective documenting the dispossession (and displacement) of San Francisco Bay Area residents in the wake of the Tech Boom 2.0."[38]

Using interactive maps, oral histories, films, murals, and events, the project highlights the "new entanglements of global capital, real estate, high tech, and political economy" in San Francisco, the most inflated and unequitable housing market in the United States. But more than developing tools and making counterdata on housing available, the Anti-Eviction Mapping Project is creating a community. Indeed, its participants aim for "collective resistance and movement building." The project, in short, uses data as handles for bringing in new members to the movement; it is "making data with community, not for community." One of its most prominent visual representations is a map of evictions in the city resulting from the Ellis Act, a state law with the following stated effects:

> Landlords have the right to evict tenants in order to "go out of business." All units in the building must be cleared of all tenants—no one can be singled out. Most often it is used to convert to condos or group-owned tenancy-in-common flats. Once a building becomes a condo it is exempt from Rent Control, regardless of the age of the building, and even if a unit owner subsequently rents to a long-term tenant.[39]

The Ellis Act visualization reveals that "no-fault" evictions are happening across San Francisco using data from a rent board, which shows that some buildings have "gone out of business" multiple times. While this is not illegal, it is a clear abuse of the law. Furthermore, units in the same buildings are being emptied and readied for

6.7

The Anti-Eviction Mapping Project, a data-driven movement. Courtesy of the author, Erin McElroy.

A Dirty Dozen Eviction
City of Change
Fighting for San Francisco's Vanishing Communities
THE BLACK WOMAN IS GOD
Ellis Act
2120
Oakland Demographic Shifts
Black Population
Section 8
390,724
169,710
90,649
412 GREEN STREET
Anti-Eviction Mapping Project
Rutele Evacuarilor Spre Strada Cantonului (1996-2016)
Leontina
1,653
Narratives of Displacement, Anti-Eviction Mapping
EVICTION DEFENSE COLLABORATIVE
Lost and Endangered Art/Community Spaces, Bay Area

shorter-term as well as higher-return vacation rentals on the home share web services Airbnb and VRBO—an action that the Anti-Eviction Mapping Project authors deem illegal. Abuses of the Ellis Act by landlords are only further exacerbating the problem of an existing real estate bubble.

The project calls attention to eviction practices, and invites audiences to fill out a survey and lend their own story to the map. Beyond that, it asks audiences to boycott landlords who profit from abuse of the Ellis Act and offers tools to look up a property's eviction history. Thus, the Anti-Eviction Mapping Project seeks not to extract information from victims of eviction but instead connect those who contribute data and draw them into an organized force.

Adapting such a model means engaging the keepers of data—those who make, manage, or use data—as well as data's subjects.[40] If we seek to create communities around data, we should approach keepers and subjects first, by engaging them in a process of participation, not after the project is complete, at which time all we can do is conduct an anodyne evaluation.[41] We should enlist keepers and subjects in understanding how to interpret the data. Beyond that, we should seek to use data visualizations to actively connect them to one another and other interested audiences. Finally, we should ask what the keepers and/or subjects need. How would they want to stay connected with the data after dissemination? How can the project be taken over and managed by those for whom the stakes are highest? What constitutes a successful project in their eyes?

CONCLUSION

The lessons of this chapter—look at the data setting, not just the data set; make place part of data presentation; take a comparative approach; collect counterdata; create interfaces that cause friction; and use data to build relationships—are not rigid prescriptions for practice. Rather, they are potentials: six ways of practicing critical reflection, best illustrated by models from the world of design, broadly construed.

What makes these projects designerly? Beyond their inventiveness, often characterized as the basis of design, these models demonstrate a commitment to examining new aesthetic sensibilities.[42] The abstract, standardized geometries of most data visualizations—their regular shapes, uniform lines, and limited palettes—frequently normalize and obfuscate the local aspects of data. Data are inherently messy. In response, conventional visualizations are generally reductive or even simplistic.

The projects in this chapter intentionally keep data strange. They don't embrace standardized visual languages. For instance, *Bear 71* immerses one in a pulsating virtual landscape layered with media. *A Sort of Joy* is a live performance. *Dear Data* toys with notation. Grassroots Mapping produces higher-than-normal resolution images. "Slave Revolt in Jamaica" employs graphical depictions from another time. The Anti-Eviction Mapping Project is the most conventional, but it is the least focused on merely showing data; the project is the community, which isn't so easily represented. Together these projects challenge us to experience data in unusual ways by not treating their presentation in a generic manner.

In my early days as a design student, I learned that models are easier to remember and adapt than principles. I offer not only models here but also interpretations: ways of seeing and talking about designs with data. Many of these models are already widely known. Yet if I can influence the way that they are understood and invoked in teaching or practice, that would be a measure of success for the book.

Although these models are more likely to be used in project-based learning, they are not specifically meant for designers. Many have noted that artists can act as the antennae for societies. The same can be said for the experimental data practitioners featured in this chapter. Their projects are feelers for unfamiliar ways of engaging with data. They illuminate new paths forward, so that the rest of us may follow.

7

LOCAL ENDS

BEYOND DATA SETS

Understanding data requires more than access to a spreadsheet. All data are entangled with places, institutions, processes, and people that fundamentally shape their significance and use. If we haven't understood the data's setting, we haven't understood the data. Over the course of four cases and six chapters, I have sought to impart this deceptively simple message along with its implications for scholarship and practice.

The case of the Arnold Arboretum illustrates the complex local attachments that data hold, suggesting that place should be an important consideration in data presentation. The DPLA reveals the challenges that can arise when data from different settings are brought into dialogue. Understanding data infrastructures means taking a comparative approach. Looking at the news as a source demonstrates that data can scarcely be separated from the algorithms used to process them. Instead of trying to differentiate the substance of data from their activation through algorithms, we should learn to see the two as part of a data system. Unpacking data requires intimate knowledge of the algorithms that shape them, and vice versa. Finally, Zillow illuminates the broader implications of interfaces: those visual, discursive, and procedural settings that shape our use of data.

Building on these lessons, the previous chapter introduces a variety of localized models intended to put the book's principles into practice through existing and accessible design examples. Such examples model the means of working with data locally, but not necessarily the broader goals in doing so.

In seeking to understand data locally, what kinds of outcomes might we hope for? Thus far, I have only hinted at the answers. Now I would like to make those aims more explicit. What does it mean to successfully put data to use in the service of local ends? As I mentioned at the outset of the book, data seem useful in the first instance because they hold the promise of insight at a distance. Yet being mindful of the local contingencies of data—both where data are made and where they are used—can reveal other benefits. Below I reflect on four commonplace goals for data: orientation, access, analysis, and optimization. Then I consider how such ambitions might be reconsidered to account for additional local ends: place making, restraint, reflexivity, and contestation.

Orientation and Place Making

Data can be important tools for orientation in complex environments. At the Arnold Arboretum, where more than seventy thousand trees, vines, and shrubs have lived since its establishment in 1872, visitors cannot hope to know the extent of the place through experience alone. Data-embossed tags on each plant specimen turn the arboretum into an inhabitable catalog for species that are not otherwise known in the surrounding region. But these same data have another role. Seeing which specifications tags hold, how they are organized, and even which plants are not tagged at all tells visitors much about the institution itself: a place in which data are deeply enmeshed in its history, materiality, and culture. Data are not just representative of the collected specimens of the arboretum; they are an integral part of the way the place works.

As another pressing example, consider the increasingly widespread notion of the smart city: a place in which all the elements of civic life are potentially mediated through data. Michael Totty reports in the *Wall Street Journal* that, "Whether it's making it easier for residents to find parking places, or guiding health inspectors to high-risk restaurants or giving smoke alarms to the households that are most likely to suffer fatal fires, big-data technologies are beginning to transform the way cities work."[1]

The smart city is heralded as a potentially seamless experience in which place and data converge to meet the needs of city residents as well as reduce administrative costs through preventive maintenance. But early efforts to develop smart city infrastructures are only beginning to grapple with the question of how the phenomenon might materialize in different local ways. We must put aside the rhetoric of digital universalism to ask, How will divergent practices of data collection and use come to characterize different manifestations of the smart city? Some places might encourage grassroots organizing and activism around data, as in the case of the Anti-Eviction Mapping Project introduced in chapter 6. Others may follow more centralized models of surveillance and control. The smart city is not one place.

Access and Restraint

Data act as bridges to large collections of digital resources that are held remotely or in distributed locations. The DPLA, for instance, brings together resources from contributing collections across the United States. Data provide access, but not in a homogeneous way. A local perspective can help users of composite collections to better understand the limitations of knowledge gathered through data.

Another significant reminder of the limits of aggregate data was the 2016 US presidential election. As sociologist of science and technology Anne Pollock and I explain in a recent essay, waves of polling data were brought together from every state in the country.[2] The flood of data seemed to indicate, overwhelmingly, that Hillary Clinton would triumph over Donald Trump. Instead, voting on November 8 produced the opposite outcome. This event fostered increased skepticism over the explanatory power of big data. And yet in the immediate wake of the election, we observed widespread efforts to redeem data. Pollsters and journalists, the producers and disseminators of election data, have responded to the apparent limitations of their polls by reworking their approaches to data aggregation and restraining their projections as opposed to rejecting polling's relevance altogether.

Analysis and Reflexivity

Data are useful not only to access large collections of media but to analyze them as well. Archives like NewsScape, described in chapter 4, can be used in conjunction with algorithms for NLP to help analysts "extract" information from the news and explore it for salient patterns. But those algorithms are not simply neutral tools to be applied to any source. They are historically and materially local, because of the data on which they were tested and even trained. Acknowledging the local conditions of such tools can help us work more reflexively, with a concrete understanding of the processes and even people—like those invisibly at work behind algorithms—that make analysis possible.

Another domain where reflexivity would benefit the application of NLP is "search." Once an experimental research project in artificial intelligence, search is now a commonplace form of automation that we use every day on Google's home page. Journalists are on the forefront of this issue, and calling public attention to how search can work to "replicate and deepen the biases" in society.[3] Carole Cadwalladr of the *Guardian*, for instance, has investigated Google's autocomplete function, which attempts to finish your search query for you, using data on common searches related to yours—and even preemptively displaying the results. She reports typing "a-r-e" and "j-e-w-s," and seeing Google complete her entry as "white," "Christian," or "evil." Furthermore, she found that typing "d-o" and "b-l-a-c-k-s" can autocomplete in Google as "commit more crimes?" When questioned by Cadwalladr about these obviously racist results, Google responded by deferring responsibility:

> Our search results are a reflection of the content across the web. This means that sometimes unpleasant portrayals of sensitive subject matter online can affect what search results appear for a given query. These results don't reflect Google's own opinions or beliefs—as a company, we strongly value a diversity of perspectives, ideas and cultures.[4]

In seeking to "organize the world's information," though, Google is anything but neutral.[5] In a recent book, information studies scholar Safiya Noble calls the company to task for creating "algorithms of oppression" that reinforce racism and sexism, particularly in search results for black girls and women.

> We have to ask what is lost, who is harmed, and what should be forgotten with the embrace of artificial intelligence in decision making. It is of no collective social benefit to organize information resources on the web through processes that solidify inequality and marginalization.[6]

Moreover, the purveyors of fake news and other forms of clickbait are explicitly working to game existing algorithms by discovering the "tricks that will move them up Google's PageRank system."[7] Learning how algorithms work in coordination with existing data can help us reflexivity attend to and take responsibility for (rather than grudgingly accept) the problems inherent in pervasive data-driven services.[8]

Optimization and Contestation

Finally, data can enable the optimization of systems that are otherwise too large or complex to configure by hand. Market-based platforms like Zillow promise to bring increased transparency and thus efficiency to the housing market by allowing buyers, sellers, and realtors to not only access real estate data but add or update data about individual homes too. Optimization, however, is only possible when the context is fixed and outcomes are agreed on. In housing, as in many areas of public life, context is always contested. Understanding how interfaces like Zillow's work to establish an operational context for data can prompt us to question the status quo implicit in all optimization efforts.[9]

Another example of the use of optimization that might be similarly contested is the "gig" economy, where employers in fast-food services, among others, use data-driven management practices to make the most efficient use of their workforce. Utilizing such techniques, businesses seek to avoid overstaffing. But the resulting work conditions for "on-demand" employees—unpredictable hours, few or no benefits, and little job security—only make sense if you understand them within the context of profit maximization.[10] How might the data currently used to optimize gig work be harnessed instead to contest unsustainable labor conditions and perhaps enhance collective bargaining between employers and labor unions?

Each of these local ends—place making, restraint, reflexivity, and contestation—are efforts to rethink the now-commonplace roles for data as universal tools for orientation, access, analysis, and optimization. Local ends require a new ethics of data work tied to local viewpoints. Using data effectively cannot simply be about securing personal benefits. Understanding data is increasingly a social and civic project.

GUIDING FUTURE RESEARCH

Accepting the message of this book that all data are local leaves us with a significant challenge: How can we make data broadly accessible while acknowledging their attachments? A general ethos of "open data" has pervaded government, academic, and industry approaches in recent years.[11] But what good are open data if they cannot be understood by outside audiences? Data made in civic and cultural institutions, exemplified by the cases in this book, are not often made in the first instance for public use. Rather, those data are part of systems for curation and management originally defined in the nineteenth century; they are expert tools, designed for use within complex knowledge systems. In order to embrace the locality of data while also making their home institutions relevant today, we must reconsider what we mean by *open data*.

One way of doing so, and the departing recommendation of this book, is to establish a new genre of open data guides, which can introduce potential users to data settings and not just data sets. What might local guides for open data consist of, who will make them, and how will they affect data use? This is a pressing research question for all of us.[12]

We might think of a local guide for open data as related to, yet more reflexive than, a traditional data user guide or manual. Such materials sometimes accompany data sets created by government institutions or organizations that hold themselves to high expectations for transparency.[13] A user guide is typically created by those who make data as a means of helping others put those data to good use.[14] It might offer a codebook, which explains each of the fields and values in the data, as well as information about their purposes, processes, and potential contexts of use. For example, the Western Pennsylvania Regional Data Center has produced a sample template for creating a data user guide with eight distinct sections: original purpose and application; history, standards, and formats; organizational context; workflow; things to know about the data (including limitations); current applications; field values; and sources and acknowledgments.[15]

This template is a productive starting point for data-creating institutions that want to make their data more accessible. But such user guides can fall short in a number of ways. First, user guides frequently do not address and might even intentionally obscure the answers to difficult ethical questions around data in order to protect their collecting institutions. Second, user guides are often fixed, while the data and regulations around their use change over time. Third, understanding data must be immersive. It requires local reading in dialogue with data experts and subjects as well as hands-on work such as data visualization. Fourth, different users have an additional burden of trying to understand their own standpoints and those of their particular audiences.

As I have explored the data featured in this book, I have also developed an awareness of my own specific subject position: a technically adept user with the time, resources, and status necessary to access complex as well as sometimes-shrouded, if ostensibly open, data settings. I do not presume to know what other, less privileged users might need to facilitate access for their own ends. For all the reasons above, I imagine that local guides for open data would need to diverge from existing templates in a number of important ways.

Beyond establishing the contents of such guides, we must consider their social context. Making local guides means moving beyond the user data dyad; there is more than one subject position for a data user. A number of users and even nonusers might need to be accounted for. Moreover, local guides need not be made by insiders: those who make the data. Outsiders, especially students, can learn from the practice of creating such local guides as a means of coming to terms with what, beyond a spreadsheet, comma-separated values file, database, or application programming interface, is necessary for understanding data.

One of the additional challenges in developing local guides for open data will be establishing a local ethics of data use. When it comes to working with data, our ends are only just if they are arrived at justly. A number of authors have written about the ethics of data involving issues of persuasion, privacy, security, and even exploitation.[16] While these issues are of utmost significance, I would ask, What unanswered ethical questions do the cases in this book suggest? For example, I expect local ethics to learn from the ethics of care. Given that data are never raw, Geoffrey Bowker tells us, they need to be cooked "with care."[17] Unfortunately, there is no formula for doing so.

An ethics of care does not presuppose a set of universal rules for treating data and associated humans (and nonhumans) ethically but rather implies maintaining relationships with them. Maria Puig de la Bellacasa writes that "we need to ask, 'how to care' in each situation."[18] Furthermore, we need to understand who is empowered by acts of caring. As the Grassroots Mapping project introduced in chapter 6 illuminates, centralized institutions, such as the World Bank, often collect data in ways that reproduce colonialist patterns of external representation and control, veiled behind the discourse of care.

A strict ethics template might lead us to another form of digital universalism or one-size-fits-all solution. We are beginning to see this in domains like the design of so-called smart cities where ethics are an obvious requirement.[19] Consider, as mentioned previously, that few city officials are thinking about what *smart* should mean in their particular locality. How will smartness address their city's local social, economic, and geographic advantages as well as challenges in ethical ways? Chapter 5 illustrates the problems of simply making civic data open—one of the defining steps toward creating the generic smart city. Publicly accessible data, as I explain in the case of Zillow, can be placed in visual, discursive, and algorithmic contexts that undermine their original role as resources for the public good. What are the local effects of Zillow on neighborhoods in Atlanta, where rampant economic speculation is displacing historically black communities at a rapid pace? What ethical obligations does Zillow have to those communities, and can those obligations be simply generalized for the entire country? Using data ethically is a local problem, which requires attention to the differences among data settings and how they might change over time, necessitating continued maintenance and adjustment. Thus a local ethic also implies evolving roles for those who act as intermediaries and stewards for data. Understanding those identities and their future potential is a necessary component of such research too.

Below I propose first steps toward creating local guides to open data in five parts, which build on the findings in this book. This sequence might be particularly useful for students, scholars, and practitioners who frequently encounter data sets that are new to them. That said, it is a provocation to further research rather than another template to be followed exactly. For we do not yet know what kind of guidance will work, where, and for whom.

Step 1: Read

Acquire a human-readable version of the data set. Try to identify at least two kinds of data: one that seems typical, and one that is surprising or confusing. Consider the data format(s). What can we learn about their history? Where else are they used? What other formats are used for similar data? Question why the data looks the way that it does.

Step 2: Inquire

Establish a rapport with a diverse group of informants who are local experts on the data set: data collectors, data analysts, or data subjects (someone who the data represents). Ask them about the provenance and purpose(s) of the data. Ask them about known patterns. Also ask them to identify problems: errors, absences, and limitations that can illuminate the context(s) of collection. With their help, create an accessible codebook, explaining the various data fields and their structure, along with your own code of ethics for working with the data set.

Step 3: Represent

Use a simple visualization technique, such as a scatterplot, line graph, timeline, map, tree map, or network diagram, to confirm or contest a known pattern in the data, first revealed by the informants. Return to the informants with any new questions prompted by the visualization.

Step 4: Unfold

Work with the informants to create a diagram of the collection, normalization, maintenance, and distribution processes used with the data set. Learn about any specialized algorithms used on or developed along with the data.

Step 5: Contextualize

Identify and analyze several contexts of use for the data. Who is using these data and what claims are they making? What visual, discursive, or algorithmic tools are they employing? What ethical issues do these uses present? What friction do they "kill" or potentially kindle?

Learning to venture into local knowledge systems has long been the role of an ethnographer: one who documents the practices of different cultures. But today we encounter new and confounding knowledge systems in each new data setting. While not all of us will become ethnographers in the traditional sense, we can learn to take an "ethnographic stance" when encountering unfamiliar sources by making a commitment to contextualization.[20] Where are data made and by whom? What assumptions and values do data carry? How are data entangled with otherwise-invisible processes? What are the contexts in which we confront data? Such questions can help us establish new expectations of data as well as our relationships with both their keepers and subjects.

We increasingly live in societies driven by data and the accompanying promise of connectivity. But as this book has sought to demonstrate, connections based solely on data are relatively weak. If I can offer one takeaway message to the reader, it is this: treat data as a point of contact, a landing, an opportunity to get closer, to learn to care about a subject, or the people and places beyond data. Do not mistake the availability of data as permission to remain at a distance.

NOTES

PREFACE

1. The renovation was covered in the *New York Times*. Cotter, "The Met Reimagines the American Story."
2. For an introduction to the concepts of "back stage" and "front stage," see Goffman, *The Presentation of Self in Everyday Life*.
3. Ibid.
4. For an introduction to sociotechnical perspectives, see Bijker, *Of Bicycles, Bakelites, and Bulbs*.
5. Galison, "Limits of Localism."
6. Turkle, *Simulation and Its Discontents*; Loukissas, *Co-Designers*.
7. Geertz, *The Interpretation of Cultures*.
8. David Small first proposed this idea.
9. In recent years, the annual meeting of the Society for the Social Studies of Science and the iConference have both incorporated sessions on "making" and "doing," which are other terms for "design." For more on the possible relations between design and science, technology, and society, see Vertesi et al., "Engaging, Designing, and Making Digital Systems."
10. I see the design of such systems in the broad terms laid out by Herbert Simon many years ago: "Everyone designs who devises courses of action aimed at changing existing situations into preferred ones." Since Simon wrote this in 1969, the term *design* has continued to gain ground, to a degree that might have surprised even him. Simon, *The Sciences of the Artificial*, 111.

INTRODUCTION

1. *Mega meta collection* is a term first used by Dan Cohen to describe "collections of collections" like the US Library of Congress's Global Gateway project. See LeFurgy, "Digging into a Slice of Digital History."
2. In recent years, researchers in data studies have sought to illuminate the local conditions in which data are created. For an early study, see Bowker and Star, *Sorting Things Out*. On reused data, see Zimmerman, "New Knowledge from Old Data." On aggregated data, see Edwards, *A Vast Machine*. On exchanged data, see Vertesi and Dourish, "The Value of Data."
3. University of California at Berkeley, College of Environmental Design, "The Data Made Me Do It."
4. Kleinman, "Artificial Intelligence." For a more in-depth academic exploration of this subject, see Noble, *Algorithms of Oppression*. Strauss, "34 Problems with Standardized Tests." For a more extensive treatment of this topic, see Muller, *The Tyranny of Metrics*. Confessore and Wakabayashi, "How Russia Harvested American Rage to Reshape U.S. Politics." For an academic exploration of this subject, see Boczkowski and Papacharissi, *Trump and the Media*.
5. In statistical analysis, p-value (probability value) is a number between 0 and 1. Results are typically considered statistically significant when p-value < 0.05. This means the chances that the same results could occur at random is less than 1 in 20. Simmons, Nelson, and Simonsohn, "False-Positive Psychology."

6. In the 1980s, Sherry Turkle documented the tension between "doing" and "doubting" around data that accompanied early efforts by scientists at MIT to incorporate computer simulations in their labs and classrooms. One physics professor explained that the machines changed "what it means to do a physical measurement." See Turkle, *Simulation and Its Discontents*, 36. For an even broader exploration of the history of these debates, see Porter, *Trust in Numbers*.
7. Haraway, "Situated Knowledges," 590; Nagel, *The View from Nowhere*.
8. For influential works on privacy within social media, see boyd, *It's Complicated*, and Marwick and boyd, "Networked Privacy."
9. Finn, *What Algorithms Want*.
10. Consider JavaScript Object Notation, which is an increasingly used standard that brings together collections thinking and object orientedness in order to serve a broad array of online services.
11. As Lewis writes, "The price of the end product was driven by the ratings assigned to it by the models used by Moody's and S&P. The inner workings of these models were, officially, a secret: Moody's and S&P claimed they were impossible to game. But everyone on Wall Street knew that the people who ran the models were ripe for exploitation." Lewis, *The Big Short*, 99.
12. Ibid., 6.
13. Ibid., 99.
14. Loukissas and Mindell, "Visual Apollo." For a full analysis of human-machine relationships during the first lunar landing, see Mindell, *Digital Apollo*.
15. Some of the findings of this study were synthesized into a conference paper. Loukissas et al., "Redesigning Postoperative Review."
16. For more on the dominance of US models of computing, see Chan, *Networking Peripheries*.
17. Ortner, *Anthropology and Social Theory*, 42.
18. To learn more about this history of data, see Gitelman, *"Raw Data" Is an Oxymoron*. Lisa Gitelman's book, though, does not deliver the necessary tools for scholarship and practice in the present. Meanwhile, Christine Borgman's work seeks to thoroughly prepare us for the challenges of data today, but it does so only for academic fields. Borgman, *Big Data, Little Data, No Data*.
19. This book is informed by extended empirical fieldwork, including interviews, observations, and workshops conducted over the course of the past six years. In preparing to write this book, I undertook the following research. I conducted fieldwork between 2012 and 2014 with faculty and staff at the Arnold Arboretum to inform chapter 2. For chapter 3, I engaged staff at the DPLA as well as contributing institutions (libraries, museums, and archives) between 2012 and 2015. Between 2014 and 2017, I worked with practitioners and researchers within broadcast journalism to inform chapter 4. From 2015 to 2017, I studied the practices of real estate agents, planners, tax assessors, developers, and affordable housing advocates in Atlanta, Georgia, to inform chapter 5. Many of the subjects interviewed or observed in my fieldwork are not mentioned directly. Subjects are only named with permission.
20. I make use of a combination of local reading techniques. This hybrid model of analysis owes much to developments in close and distant reading as methods of interrogating texts in liter-

ary and cultural studies. When used as a method of analysis for literary texts, Jonathan Culler explains that close readings attend to "how meaning is produced or conveyed. Culler, "The Closeness of Close Reading," 22. Paradoxically, distant reading aims not to read. Instead, the latter technique, pioneered in literature by Franco Moretti, aspires to "generate an abstract view by shifting from observing textual content to visualizing global features of a single or of multiple text(s)." Jaenicke and Franzini, "On Close and Distant Reading in Digital Humanities," 2.

21. Geertz, "Local Knowledge and Its Limits."
22. For more on speculative approaches, see Dunne and Raby, *Speculative Everything*.
23. The visualizations in this book represent a small fraction of the total number developed during the course of my research. Each is custom created from computer code, primarily using Java or JavaScript, and sometimes D3 (Data-Driven Documents), a JavaScript library created by Mike Bostock and others. Many of the visualizations were made in collaboration with two graduate students of mine, Krystelle Denis (Harvard University) and Peter Polack (Georgia Tech). Often, I created an initial code sketch, and then my students elaborated on it. We did not rely on premade visualization templates or software. Yet I have left out the vast majority of the visualizations we created, for they were frequently rudimentary and of limited value to the reader. I have included only examples that help me make my points, and even then, only the last iterations.
24. Data visualization is significantly informed and enabled by, but more publicly oriented than, the subfield of computer science called information visualization. For an extensive history of these practices, see Drucker, *Graphesis*. My own particular approach to visual design and communication is shaped by my education in manual drafting at Cornell University's Department of Architecture in the 1990s.
25. Discussions of what makes an effective data visualization are often kept separate from questions about the origins of data as well as their local stakes. On efficiency, see Tufte, *The Visual Display of Quantitative Information*. On memorability, see Borkin et al., "What Makes a Visualization Memorable?" On elegance, see Kirk, *Data Visualization*. "Data graphics," writes Tufte, "are instruments for reasoning about quantitative information." Tufte, *The Visual Display of Quantitative Information*, 9. "Whenever we analyze data," explains Ben Fry, "our goal is to highlight its features in order of their importance, reveal patterns, and simultaneously show features that exist across multiple dimensions." Fry, *Visualizing Data*, 1. "Data visualization," instructs Andy Kirk, is "the representation and presentation of data to facilitate understanding." Kirk, *Data Visualization*, 19. For one of the only books to embed a critique within data visualization, see Kurgan, *Close Up at a Distance*. Her monograph, however, is more a collection of creative work than a theoretical treatment of data.
26. This term was first used in Dalton and Thatcher, "What Does a Critical Data Studies Look Like, and Why Do We Care?"
27. Kitchin and Lauriault, "Towards Critical Data Studies."
28. Bogost, *Persuasive Games*; Elmborg, "Critical Information Literacy"; Ratto, "Critical Making"; Dunne, *Hertzian Tales*.
29. Bellacasa, *Matters of Care*.
30. Maria Puig de la Bellacasa writes "understanding caring as something we do extends a vision of care as an ethically and politically charged practice, one that has been at the forefront of

feminist concern with devalued labors." Bellacasa, "Matters of Care in Technoscience," 90. The need for care is widespread, but not simple. David Ribes explains that care can go wrong, specifically in practices of maintenance and repair: "Whether drawing attention to them, valuing them, or conceptually repopulating them with human work, care—practice or history—cannot be an end in and of itself if it means failing to ask the questions: repair and maintenance of what, serving whose interests, and at the expense of what people?" Ribes, "The Rub and Chafe of Maintenance and Repair."

31. Negroponte, *Being Digital*, 165.
32. Negroponte, "One Laptop per Child."
33. Robertson, "OLPC's $100 Laptop Was Going to Change the World."
34. Ananny and Winters, "Designing for Development," 107.
35. Berners-Lee et al., "The World-Wide Web."
36. Many eminent technologists, such as Ted Nelson, have criticized the founding assumptions of the web, which is now the dominant standard for the internet. See Nelson, "Complex Information Processing."
37. You do not go to a website. It is downloaded and displayed on your machine. Many people do not consciously realize that by the time they've seen a website, it is already stored on their computer and it will be indefinitely. Furthermore, depending on your network connection, hardware, software, and settings—not to mention any laws of your country that censor its content—the web page may appear quite differently that it does in other places.
38. Chan, *Networking Peripheries*, 7.
39. See Taub and Fisher, "Where Countries Are Tinderboxes and Facebook Is a Match."
40. For further reading on free market ideology, see Graeber, *Toward an Anthropological Theory of Value*.
41. For a detailed explanation of computing at the periphery in Peru, see Chan, *Networking Peripheries*. For more on how data can become enlisted in colonialism, see Thatcher, O'Sullivan, and Mahmoudi, "Data Colonialism through Accumulation by Dispossession."

CHAPTER 1

1. Chalabi, "'Data Is' vs. 'Data Are.'"
2. For more about how the term *data* can be used, see the *Associated Press Stylebook* or *Chicago Manual of Style*.
3. Rob Kitchin explains that the use of capta instead of data was also called for by H. E. Jensen in 1950. See Kitchin, *The Data Revolution*.
4. Dourish, *The Stuff of Bits*, 107.
5. Ibid., 4.
6. Crampton, *Mapping*.
7. Bowker and Star, *Sorting Things Out*; Gitelman, *"Raw Data" Is an Oxymoron*; Dourish, *The Stuff of Bits*.
8. Latour, "Drawing Things Together," 26.

9. Ibid.

10. See Latour and Woolgar, *Laboratory Life*; Lynch, *Art and Artifact in Laboratory Science*; Cetina, *Epistemic Cultures*; Keller, *Making Sense of Life*; Star and Griesemer, "Institutional Ecology, 'Translations' and Boundary Objects"; Bowker and Star, *Sorting Things Out*; Vertesi and Dourish, "The Value of Data"; Edwards, *A Vast Machine*.

11. Borgman, *Big Data, Little Data, No Data*.

12. Kitchen extends our understanding of data's diversity beyond the academic world and calls for more specific empirical studies of diverse data cultures. Kitchin, *The Data Revolution*, 4.

13. Recounted by the author from a public lecture by Culler at Cornell University in 1998.

14. If the examples used in this book seem accessible, outside their originally intended settings, it is only because of the standardization and coordination work that has been done by their keepers, who have prepared their data to "jump contexts" from one site of use to another. See Downey, "Making Media Work."

15. Kitchin and Lauriault, "Towards Critical Data Studies."

16. Kitchin, "Big Data, New Epistemologies, and Paradigm Shifts."

17. Kitchin and McArdle, "What Makes Big Data, Big Data?"

18. Crawford, "The Anxieties of Big Data."

19. Ibid.

20. Edwards, *A Vast Machine*.

21. Anderson, "The End of Theory."

22. See Borgman, *Big Data, Little Data, No Data*, 17. The early relation between data and faith has not entirely dissipated. Issues of purity still pervade the handling of data.

23. Buckland, "Information as Thing"; Borgman, *Big Data, Little Data, No Data*.

24. Borgman, *Big Data, Little Data, No Data*, 28.

25. Ibid.

26. Ribes and Finholt, "The Long Now of Technology Infrastructure."

27. See Posner and Klein, "Data as Media"; Lievrouw, "Materiality and Media in Communication and Technology Studies."

28. Marshall McLuhan writes, "This is merely to say that the personal and social consequences of any medium—that is, of any extension of ourselves—result from the new scale that is introduced into our affairs by each extension of ourselves, or by any new technology." McLuhan, *Understanding Media*, 9.

29. Manovich, *The Language of New Media*, 225.

30. McLuhan, *Understanding Media*, 3.

31. Gillespie, Boczkowski, and Foot, *Media Technologies*.

32. In everyday terms, *local* means "relating to a particular region or part, or to each of any number of these." *Oxford English Dictionary*, s.v. "Local, adj. 2," accessed December 4, 2017, http://www.oed.com/viewdictionaryentry/Entry/11125.

33. Geertz, *Local Knowledge*.

34. Geertz, *Available Light*, 140.

35. See Harding, "Is Science Multicultural?"; Wyer et al., *Women, Science, and Technology*; Barad, *Meeting the Universe Halfway*. For more on how feminist theory has been used to understand data and data visualization, see D'Ignazio and Klein, "Feminist Data Visualization."

36. See Harding, "Is Science Multicultural?," 55. As an example, Harding notes that "Western conceptions of laws of nature drew on both Judeo-Christian religious beliefs and the increasing familiarity in early modern Europe with centralized royal authority, with royal absolutism." Meanwhile, Chinese science regarded nature as a "web of relationships without a weaver," not an empire ruled by divine laws. See Ibid., 316. For more on knowledge systems and their relation to local knowledge, see Turnbull, "Local Knowledge and Comparative Scientific Traditions"; Watson-Verran and Turnbull, "Knowledge Systems as Assemblages of Local Knowledge."

37. More recently, for instance, John Law and Annemarie Mol describe how scientific facts are "brought down to earth" through early laboratory studies, such as at an institute for biological studies in San Diego or plant protein research facility in Berkeley, both in California. Law and Mol, "Situating Technoscience," 610.

38. See Star and Griesemer, "Institutional Ecology, 'Translations,' and Boundary Objects"; Edwards, *A Vast Machine*; Vertesi and Dourish, "The Value of Data."

39. Haraway, "Situated Knowledges"; Suchman, *Human-Machine Reconfigurations*.

40. Cresswell, *In Place/Out of Place*, 3, 10.

41. Frampton, "Towards a Critical Regionalism."

42. Gibson, *Neuromancer.*

43. See Benedikt, *Cyberspace*; Mitchell, *City of Bits*; Castells, *The Information Age*. These authors have overturned a long history of social theories on the local. For studies of gemeinschaft and gesellschaft, see Tonnies and Loomis, *Community and Society*. On the distinction between particularism and universalism, see Parsons, *The Social System*.

44. Mitchell, *Me++*, 14.

45. Castells, *The Information Age*.

46. More recently, scholarship has shown the way that software and space mutually inform one another. See Kitchin and Dodge, *Code/Space*. We have yet to come to terms with this relationship at the level of data, however.

47. Ali, *Media Localism*, 5.

48. Kovach and Rosenstiel, *The Elements of Journalism*, 160.

49. Borgman, *Big Data, Little Data, No Data*.

50. Bowker and Star, *Sorting Things Out*.

51. Wagner, "The Americans Our Government Won't Count."

52. Bowker, "Biodiversity Datadiversity"; Gitelman, *"Raw Data" Is an Oxymoron*.

53. Geertz, *Local Knowledge*, 129.

54. Gieryn, "A Space for Place in Sociology"; Edwards, *A Vast Machine*.

55. For further reading on how nested locality can work, see Bratton, *The Stack*.

56. Star and Ruhleder, "Step toward an Ecology of Infrastructure," 114.
57. Edwards, *A Vast Machine*, 3.
58. See Galison, "Limits of Localism."
59. Pariser, *The Filter Bubble*.
60. Leurs, "Feminist and Postcolonial Data Analysis"; Houston and Jackson, "Caring for the Next Billion Mobile Handsets."
61. Tronto, *Moral Boundaries*, 103.
62. From a DPLA meeting attended by the author in 2013 in Cambridge, Massachusetts. See also Battles and Loukissas, "Data Artefacts."
63. Battles and Loukissas, "Data Artefacts."
64. Bellacasa, *Matters of Care*.
65. For more about gendered approaches to technical learning, see Turkle and Papert, "Epistemological Pluralism."

CHAPTER 2

1. Data from the Arnold Arboretum are accessible online at https://www.arboretum.harvard.edu/plants/data-resources.
2. At the time, I was a Media Arts Fellow at Harvard University and member of the research group metaLAB@Harvard, part of the Berkman Klein Center for Internet and Society. Hailing from a range of humanistic subfields, this group was united by a determination to invent new counterdisciplinary approaches to working with collections of science and cultural heritage. Those present at Bussey Brook that day included Mathew Battles, Kyle Parry, and Jessica Yurkofsky as well as our host, arboretum senior researcher Peter Del Tredici.
3. For more on urban ecology in the Northeast, see Del Tredici, *Wild Urban Plants of the Northeast*.
4. For an extended contemplation of Bussey Brook Meadow, see Battles, *Tree*.
5. "Public Lab."
6. For more on aerial photography and its history, see Haffner, *The View from Above*.
7. From an interview by the author with Michael Dosmann, 2014.
8. Note that the provenance is not a place of origin but rather the name and address of a collector.
9. From an interview by the author with Peter Del Tredici, 2014.
10. Ibid.
11. This understanding reinforces prior studies of data that show institutionalized categories to be connected to specific social groups. Star and Griesemer, "Institutional Ecology, 'Translations,' and Boundary Objects."
12. From a record in BG-Base, the Arnold Arboretum's database of plant accessions.
13. Del Tredici and Kitajima, "Finding a Replacement for the Eastern Hemlock."
14. From an interview by the author with Peter Del Tredici, 2014.

15. For a detailed exploration of this concept in citizen science, see Gabrys, Pritchard, and Barratt, "Just Good Enough Data."
16. Edwards, *A Vast Machine*.
17. See Mayernik, Batcheller, and Borgman, "How Institutional Factors Influence the Creation of Scientific Metadata." See also Edwards et al., "Science Friction."
18. Gnoli, "Metadata about What?"
19. Kitchin and Lauriault, "Towards Critical Data Studies."
20. From an interview by the author with Peter Del Tredici, 2014.
21. For place-based perspectives on science and technology, see Galison and Thompson, *The Architecture of Science*; Livingstone, *Putting Science in Its Place*; Kirschenbaum, *Mechanisms*.
22. Haraway, "Situated Knowledges," 589.
23. Cosgrove, *Geography and Vision*; Crampton, *Mapping*; Kitchin, Gleeson, and Dodge, "Unfolding Mapping Practices"; Wilson, *New Lines*.
24. See Schivelbusch, *The Railway Journey*. STS scholar Laura Forlano first connected the use of this term to Schivelbusch's work in a phone conversation with the author.
25. For further reading on visualizations as instruments for reframing, see Hall, "Critical Visualization."
26. Accessible at http://www.lifeanddeathofdata.org.
27. From an interview by the author with Michael Dosmann, 2014.
28. Ibid.
29. Battles and Loukissas, "Data Artefacts."
30. Douglas, *Purity and Danger*; Mody, "A Little Dirt Never Hurt Anyone."
31. From an interview by the author with Peter Del Tredici, 2014.
32. Ibid.
33. From an interview by the author with Kyle Port, 2014.
34. The form of this visualization was influenced by a DensityDesign project, entitled "La Lettura," completed in 2012 in collaboration with Corriere della Sera. See https://densitydesign.org/research/la-lettura/, accessed July 27, 2018.
35. Morozov, "The Rise of Data and the Death of Politics"; Anderson, "The End of Theory"; Lohr, "The Age of Big Data."
36. Kalay and Marx, "The Role of Place in Cyberspace"; Mitchell, *City of Bits*; Graham, "The End of Geography or the Explosion of Place?"; Dourish, "Re-Space-Ing Place"; Irani et al., "Postcolonial Computing."
37. Dalton and Thatcher, "What Does a Critical Data Studies Look Like, and Why Do We Care?," 6.
38. Gieryn, "A Space for Place in Sociology." John Agnew writes of places as "discrete if 'elastic' areas in which settings for the constitution of social relations are located and with which people can identify." Agnew, "Representing Space," 263.
39. Buell, *The Future of Environmental Criticism*.

40. Hacking, "A Tradition of Natural Kinds"; Kitchin and Lauriault, "Towards Critical Data Studies."
41. Rob Kitchin and Gavin McArdle define big data as high magnitude in volume (terebytes or petabytes), velocity, variety, scope, resolution, flexibility, and relations with other data sets. Kitchin and McArdle, "What Makes Big Data, Big Data?" This litany of attributes, however, accounts for only the most ambitious of contemporary practices with big data. My use of the term is more in line with the work of danah boyd and Kate Crawford, who characterize big data as a phenomenon with not only technological but also cultural dimensions. See boyd and Crawford, "Critical Questions for Big Data."
42. See Battles, *Library*.
43. Hayles, *How We Became Posthuman*.

CHAPTER 3

1. Data from the DPLA are accessible online at https://pro.dp.la/developers.
2. At the time of the Appfest, the DPLA was a project of the Berkman Klein Center for Internet and Society, an institution known for its efforts to map and remap the legal problems of the internet. The reader might be familiar with "Creative Commons," its revolutionary copyright schema currently in use across the web.
3. See http://dp.la/info/about/history.
4. For more on data infrastructures, see Kitchin, *The Data Revolution*.
5. After the first year as a nonprofit, the DPLA was well aware of the challenges of ingestion. The ingestion process exposed problems in the consistency, enrichment, and validation of the final data. As a result, these processes required significant hands-on work from DPLA staff—an approach that was presented as unsustainable at the time. See Matienzo and Rudersdorf, "The Digital Public Library of America Ingestion Ecosystem."
6. Palfrey and Gasser, *Interop*.
7. "One of our major priorities is to improve the quality of data we receive from libraries, archives, and museums. After all, it is from this data that so much else flows, including our ability to search well across collections and develop popular interfaces such as our browseable map. We plan to work extensively with our partners to get better data into this core piece of DPLA, and to refine that data as part of our valuable ingestion processing." "Digital Public Library of America Strategic Plan, 13.
8. Much has been written on the challenges of interface design; see Negroponte, *Being Digital*; Manovich, *The Language of New Media*; Galloway, *The Interface Effect*; Hookway, *Interface*.
9. "DPLA Metadata Application Profile."
10. Latour, *Science in Action*.
11. Not unlike the kind of conversation with materials that Donald Schön described as design. See Schön, *The Reflective Practitioner*. In fact, Turkle and Schön developed the notion of the "technological bricoleur" together while working on a study of MIT's Athena computing system and its implications for pedagogy at the institute. See Turkle and Schön, "The Athena Project." Also see Turkle, *Life on the Screen*.
12. The other participants in my Appfest group were Matthew Battles, Joshua Cash, Jessica Donaldson, Summer Leimart, Jim Reece, and Jeremy Throne.

13. We nonetheless went on to receive a small grant from the DPLA to develop our interface into the Library Observatory.

14. "Our survey and analysis have strengthened our conviction that multiple copies of a pre-copyright book cannot be deemed redundant based on catalog information alone." Ruotolo et al., "Book Traces @ UVA."

15. Ibid

16. Howard, "Book Lovers Record Traces of 19th-Century Readers."

17. Cited in Lewis, "ASU, MIT to Develop New Approaches to Library Print Collections."

18. "Digital Public Library of America Strategic Plan," 3.

19. From an interview by the author with Jeffrey Licht, 2013.

20. Battles and Loukissas, "Data Artefacts."

21. From a public lecture in 2015 in Cambridge, Massachusetts, by Marya McQuirter at Beautiful Data II.

22. Klein, "The Image of Absence."

23. Gusterson, *Nuclear Rites*.

24. From an interview by the author with Thomas Ma, 2013.

25. Ibid.

26. On the museum, see, for example, Star and Griesemer, "Institutional Ecology, 'Translations,' and Boundary Objects"; Battles and Loukissas, "Data Artefacts." On the laboratory, see, for example, Latour and Woolgar, *Laboratory Life*; Cetina, *Epistemic Cultures*; Lynch, *Art and Artifact in Laboratory Science*.

27. Gitelman, *"Raw Data" Is an Oxymoron*, 3.

28. Until recently, the big data phenomenon has largely been debated independently of local sites of production and use. Scholars of big data have sought to define their subjects in more general terms. On technical definitions, in terms of their "volume, velocity, and variety," see Kitchin and McArdle, "What Makes Big Data, Big Data?" On historical definitions, as a step change from past practices, see Anderson, "The End of Theory." On practical definitions, focusing on the difficulty of managing them, see Shneiderman, "The Big Picture for Big Data." On ideological definitions, by illuminating their "mythology," see boyd and Crawford, "Critical Questions for Big Data." Moreover, popular articles about big data depict them as ubiquitous tools for research and decision making across domains. See Lohr, "The Age of Big Data"; Mayer-Schoenberger and Cukier, "The Rise of Big Data."

29. Mayer-Schoenberger and Cukier, "The Rise of Big Data."

30. Bowker and Star, *Sorting Things Out*, 34.

31. See Karadkar et al., "Introduction to the Digital Public Library of America API."

32. For more on the origins of this data format, see "Dublin Core Metadata Initiative."

33. Bush, "As We May Think."

34. Ibid. Orit Halpern writes that memex "is considered an important contribution to the dream of networked, hyperlinked, and personal computing machines." See Halpern, *Beautiful Data*, 73.

35. Bush, "As We May Think."

36. Halpern, *Beautiful Data*, 73.
37. Nelson and Brand, *Computer Lib/Dream Machines*.
38. It "would certainly beat the usual file clerk," he exclaims. See Bush, "As We May Think."
39. Kitchin, *The Data Revolution*.
40. Mayer-Schoenberger and Cukier, "The Rise of Big Data."
41. Retrieved originally by Matthew Battles from Darton, "Jefferson's Taper."
42. Nye, *American Technological Sublime*.
43. boyd and Crawford, "Critical Questions for Big Data."

CHAPTER 4

1. Data from the NewsScape archive are accessible online at http://tvnews.library.ucla.edu.
2 Like several other data visualizations in this book, I created NewsSpeak in collaboration with a former Georgia Tech graduate student, Peter Polack.
3 "UCLA Library NewsScape."
4. Others include Media Cloud, based at MIT, and the Internet Archive.
5. Dourish, *The Stuff of Bits*, 4.
6. Mayer-Schoenberger and Cukier, "The Rise of Big Data."
7. For a comprehensive look at how newspapers went digital, see Boczkowski, *Digitizing the News*.
8. Gillespie, "The Relevance of Algorithms," 167.
9. Manovich, *The Language of New Media*, 223.
10. See Eubanks, *Automating Inequality*; Noble, *Algorithms of Oppression*; Finn, *What Algorithms Want*. Social research is only beginning to address NLP algorithms specifically. See Hovy and Spruit, "The Social Impact of Natural Language Processing."
11. Loukissas and Pollock, "After Big Data Failed."
12. Office of the Director of National Intelligence, "Joint DHS, ODNI, FBI Statement on Russian Malicious Cyber Activity."
13. CNBC reports that "at his first press conference since Election Day, President-elect Donald Trump answers questions about fake news and the media, while avoiding press questions from organizations (such as CNN) he considers 'terrible.'" See "Trump to CNN Reporter."
14. As CNN reports, "Executives from Facebook, Twitter and Google are set to testify before Congress on Tuesday about foreign nationals meddling in the 2016 elections." See Fiegerman and Byers, "Facebook, Twitter, Google Testify before Congress."
15. Bird, Klein, and Loper, "Preface."
16. Recent interest in machine learning has been spurred by accessible online courses such as What Is Machine Learning? taught by Andrew Ng on Coursera.
17. This is an approach explicitly suggested by Rob Kitchin and Gavin McArdle: "Rather than studying an algorithm created by others, a researcher reflects on and critically interrogates

their own experiences of translating and formulating an algorithm." See Kitchin and McArdle, "What Makes Big Data, Big Data?," 23.

18. See, for example "Stanford CoreNLP."
19. This pseudocode was written by Georgia Tech student Benjamin Sugar, based on a program collaboratively created by the author and Peter Polack.
20. Steen and Turner, "Multimodal Construction Grammar."
21. Based on an interview by the author with a professional transcriptionist.
22. One well-known example of this is the macro "Barry the bomber," which became a shorthand for Barack Obama in the United Kingdom, where transcribers had trouble pronouncing his name. (Obama was known by this nickname at Occidental College, where he was on the basketball team and excelled at long-range shots.) There is at least one known occurrence when the closed captioning read the macro rather than the US president's name, to the embarrassment of the broadcasters.
23. From an interview by the author with Sergio Goldenberg, 2016.
24. In the past, storing the news was only something done by large libraries. It was held in basements or on microfiche, rarely accessed, and typically only by serious researchers or journalists looking for a backstory.
25. Rogers, "Wall Street Journal."
26. Based on *Wall Street Journal*, "2016 Presidential Election Calendar."
27. See, for example, Nasukawa and Yi, "Sentiment Analysis"; Pang, Lee, and Vaithyanathan, "Thumbs Up?"
28. In earlier chapters, I explain the use of visualization for reframing as a kind of critical reflection. See Hall, "Critical Visualization."
29. The idea that conventional forms of reading are innately human is a bit naive. After all, the typography of modern printing, the bound form of the codex, the institutionalization of book collecting, and even the work of the artificial light bulb make the way we read its own form of human-machine interaction.
30. From an interview by the author, 2016.
31. Watzman, "Internet Archive TV News Lab."
32. Benkler, Faris, and Roberts, "Study."
33. Costanza-Chock and Rey-Mazón, "PageOneX."
34. See Snopes.
35. Albright, "Itemized Posts and Historical Engagement"; Timberg, "Russian Propaganda May Have Been Shared Hundreds of Millions of Times."
36. Fake News Challenge.
37. Jones, "Natural Language Processing."
38. As Peter Hancox writes, "Its power came from its very limited domain and any attempt to scale the system up would result in increasingly less effective systems." See Hancox, "A Brief History of NLP."

39. Boguraev et al., "The Derivation of a Grammatically Indexed Lexicon from the Longman Dictionary of Contemporary English."

40. Ralph, "Information Extraction," 2.

41. Fillmore, "Encounters with Language," 711.

42. "Natural Language Toolkit."

43. Even today, statistical NLP is still dependent on working in specific domains. Watson, the machine that won the TV gameshow *Jeopardy* over human contestants, is a modern marvel, but only when it has the right kind of data—data that are expensive to generate and not widely available.

44. Weizenbaum, "Computer Power and Human Reason," 371.

45. For an example of how to read NLP as an otherworldly means of interacting with text, see Binder, "Alien Reading."

46. For more on training algorithms, see Diakopoulos, "Algorithmic Accountability."

47. Taylor, Marcus, and Santorini, "The Penn Treebank."

48. For an explanation of the Penn Treebank, see "Wayback Machine."

49. "The Penn Treebank (PTB) project selected 2,499 stories from a three-year *Wall Street Journal* (WSJ) collection of 98,732 stories for syntactic annotation." See "Wayback Machine."

50. For the complete NLTK book, see http://www.nltk.org/book/ch05.html.

51. Marcus, Marcinkiewicz, and Santorini, "Building a Large Annotated Corpus of English," 316.

52. There are increasingly alternatives to the Penn Treebank. Today many technology companies, such as Amazon and Facebook, need their own domain-specific annotated corpora. They are hiring people specifically to help them build these.

53. Marcus, Marcinkiewicz, and Santorini, "Building a Large Annotated Corpus of English," 313.

54. At the time of its use in the formation of the Penn Treebank, when it became the model for natural language in NLTK, the *WSJ* was only published five days per week. A weekend edition was added in 2005.

55. This is more true in English than in other languages.

56. Houston, "The Mysterious Origins of Punctuation."

57. Eisenstein, "What to Do about Bad Language on the Internet"; Brock, "From the Blackhand Side."

58. Kitchin, *The Data Revolution*.

59. Gillespie, "The Relevance of Algorithms"; Seaver, "Knowing Algorithms." Samir Passi and Steven Jackson have explained this as a kind of data vision: "the ability to organize and manipulate the world with data and algorithms, while simultaneously mastering forms of discretion around why, how, and when to apply and improvise around established methods and tools in the wake of empirical diversity." Passi and Jackson, "Data Vision."

60. Manovich, *The Language of New Media*.

61. Borges, *Collected Fictions*.

62. Dourish, *The Stuff of Bits*.

63. Baudrillard, *Simulacra and Simulation*.
64. Based loosely on the notion that making media helps people make knowledge. See Piaget, *The Psychology of Intelligence*; Papert, *Mindstorms*.
65. Latour, *Reassembling the Social*; Law and Benschop, "Resisting Pictures"; Mol, *The Body Multiple*. For an example of actor-network theory applied to the news, see Turner, "Actor-Networking the News."
66. Latour wrote that "the set of statements considered too costly to modify constitute what is referred to as reality." The news, like science, is a contest to construct reality. "Scientific activity is not 'about nature,' it is a fierce fight to construct reality." See Latour and Woolgar, *Laboratory Life*, 243.
67. Data and algorithms are in a symbiotic relationship, argues Manovich. When algorithms are used to manipulate data, the more complex the data structure, the simpler the algorithm can be. But this way of thinking is reductive and overly generalized. How can we turn from claims about cultural algorithms to the exploration of algorithmic practices, and how they reshape both data and reality?
68. From an interview by the author with Sergio Goldenberg, 2016. Also see Goldenberg's work in Murray et al., "StoryLines."
69. Kovach and Rosenstiel, *The Elements of Journalism*.
70. Zuckerman, *Digital Cosmopolitans*.

CHAPTER 5

1. Data from the Fulton County Board of Tax Assessors are accessible online at http://fultonassessor.org/.
2. According to "Rise of the Corporate Landlord," "Companies like the Blackstone Group, American Homes4Rent, Colony Financial, Silver Bay, Starwood Waypoint, and American Residential have spent approximately $20 billion to purchase roughly 150,000 single family homes, nationwide, and convert them into rental properties." Also see Call, Powell, and Heck, "Blackstone."
3. Atlanta Intown neighborhoods are inside the I-285 loop.
4. Battles, *Tree*.
5. Hanan, "Home Is Where the Capital Is."
6. Immergluck, Carpenter, and Lueders, "Declines in Low-Cost Rented Housing Units in Eight Large Southeastern Cities."
7. Units costing $750 per month or less are dwindling at the rate of 5 percent per year. See ibid.
8. See Finn, *What Algorithms Want*. The phrase "interface layer" is used by designer turned venture investor Scott Belsky. Belsky, "The Interface Layer."
9. Belsky, "The Interface Layer."
10. When Adele Goldberg and Alan Kay first wrote about the "personal" computer, they were imagining an expressive tool, not an instrument for consumerism. See Kay and Goldberg, "Personal Dynamic Media." Experimental humanities scholar Ed Finn explains that companies working to create the "interface economy" establish "wrappers around existing services,

bundling, organizing, and demystifying them for a painless user experience." Finn, *What Algorithms Want*, 129.

11. See Kitchin, *The Data Revolution*; Ramirez et al., "Data Brokers"; Small, *FTC Report Examines Data Brokers*.
12. Yelp.
13. Nextdoor.
14. Uber.
15. "Zillow."
16. Galloway, *The Interface Effect*.
17. For a summary of the confusion over what counts as context in information technology design, see Seaver, "The Nice Thing about Context Is That Everyone Has It."
18. For example, this was a common refrain at the 2016 information design and visualization conference, Information+, in Vancouver, Canada.
19. See O'Neil and Schutt, *Doing Data Science*. See also popular articles such as Richtel, "How Big Data Is Playing Recruiter for Specialized Workers."
20. For more on the problem of technological platitudes, see Paul Dourish's critical studies of disembodied interaction. Dourish, *Where the Action Is*.
21. Meanwhile, there are humans, particularly in communities that do not or cannot produce their own data, who actually lack a voice in deliberations over issues of utmost importance to them.
22. As danah boyd and Kate Crawford write, "There is a value to analyzing data abstractions, yet retaining context remains critical, particularly for certain lines of inquiry. Context is hard to interpret at scale and even harder when data are reduced to fit into a model. Managing context in light of Big Data will be an ongoing challenge." boyd and Crawford, "Critical Questions for Big Data."
23. Loukissas, "Taking Big Data Apart."
24. Today, context is a topic of widespread interest in communities that develop information technologies. Researchers and designers seek to shift the burden of accounting for the relevant circumstances of interaction from humans onto computers. In the course of their development of "context-aware" systems, they have, understandably, put forward definitions of *context* that fit well within a data processing framework. Some adopt a definition that presents *context* as data not taken into account, but that can be retrieved at a later point. Other definitions suggest a more dynamic approach, treating *context* as data that change based on the system and its use. Some see it as the users' environment, while others view *context* as the environment of the computing system. Anind Dey offers an encompassing informational view: "Context is any information that can be used to characterize the situation of an entity. An entity is a person, place, or object that is considered relevant to the interaction between a user and an application, including the user and applications themselves." Dey, "Understanding and Using Context," 3. All these definitions frame *context* as a kind of data or information.
25. Dourish, "What We Talk about When We Talk about Context," 22.
26. Ibid.

27. Although not the first to use the term, design theorist Donald Norman popularized the word *affordance* for explaining the properties of interactive objects. See Norman, *The Design of Everyday Things*.
28. Geertz, *The Interpretation of Cultures*, 14.
29. See foundational work on epistemic cultures and communities. Cetina, *Epistemic Cultures*; Keller, *Making Sense of Life*.
30. This thought experiment was first described by philosopher Gilbert Ryle and later adapted by Geertz. See Geertz, *The Interpretation of Cultures*.
31. Ibid., 7.
32. Rascoff and Humphries, *Zillow Talk*, 17.
33. Or it might be a reference to a children's book by Dr. Seuss titled *There's a Wocket in My Pocket*, in which the last line is "The ZILLOW on my PILLOW always helps me fall asleep."
34. Many internet companies such as Uber and Airbnb similarly seek to intervene in a market without adhering to previously existing rules and regulations.
35. Realtor.com was the first.
36. For an overview of data brokers and their practices, see Small, *FTC Report Examines Data Brokers*.
37. From remarks by Stewart Brand at the Hackers Conference in Marin County, California, in 1984.
38. For a more in-depth history of mapping technologies, see Drucker, *Graphesis*; Crampton, *Mapping*; Kitchin, Gleeson, and Dodge, "Unfolding Mapping Practices."
39. The complex history of this phrase was the subject of a *New York Times* article. See Safire, "Location, Location, Location."
40. Robinson, *The Look of Maps*.
41. For a description of how representations produce reality, see Baudrillard, *Simulacra and Simulation*; Wood and Fels, *The Natures of Maps*.
42. See Kitchin, Gleeson, and Dodge, "Unfolding Mapping Practices," 15.
43. Sellers know that resetting the listing date can make their property more appealing. In this sense it is kind of like a digital paint job. After all, properties that have been sitting on the market for too long, like houses where the paint is peeling, look neglected and hence less desirable.
44. Much has been written about information technology and discourse that I will not reprise here. My use of the term *discourse* as a means of establishing and maintaining power relations is rooted in the work of Michel Foucault. See *The Archaeology of Knowledge: And the Discourse on Language*.
45. For the still-best reference for understanding the term *public* and its implications, see Dewey, *The Public and Its Problems*. For further reading on the concept of *publics* and its relationship to design, see LeDantec, "Infrastructuring and the Formation of Publics in Participatory Design." For more on the ethics of using big data that's publicly available, see boyd and Crawford, "Critical Questions for Big Data."

46. See https://zillow.zendesk.com/hc/en-us/articles/213218507-Where-does-Zillow-get-infor-mation-about-my-property.

47. See https://zillow.zendesk.com/hc/en-us/articles/212561938-What-is-a-property-page.

48. From an interview by the author, 2016.

49. From an interview by the author, 2016.

50. See Rascoff and Humphries, *Zillow Talk*.

51. The only exceptions to this that I have found are in low-income neighborhoods. Some multi-unit properties do not get a Zestimate. Is Zillow not able to find any records on these units? We cannot be certain. But there are some homes that Zillow's algorithm implicitly does include in its otherwise-comprehensive perspective on the market. Thanks to Juan Carlos Rodriguez for first calling attention to such properties along the Buford Highway corridor in Atlanta.

52. In some states, the Multiple Listing Service, which works directly with realtors to collect data about properties on the market, also provides Zillow with bare-bones listings. This is the case in Georgia, where First Multiple Listing Service Inc. has partnered with Zillow to allow information listed in their databases to be used.

53. See Rascoff and Humphries, *Zillow Talk*.

54. Seaver, "Knowing Algorithms," 9.

55. See http://www.zillow.com.

56. "Pricing conversations that may be started with the review of the Zestimate should ultimately be augmented with the input of opinions from local real estate professionals (agents, brokers, and appraisers)." See http://www.zillow.com/research/putting-accuracy-in-context-3255.

57. Howell, "How Accurate Is Zillow's Zestimate?"

58. Humphries, "How Accurate Is the Zestimate?"

59. Rascoff and Humphries, *Zillow Talk*, 223.

60. Humphries, "How Accurate Is the Zestimate?"

61. "In September 2012, the Zestimate was just as likely to be too low as too high; now, it is roughly twice as likely to be too low." Howell, "How Accurate Is Zillow's Zestimate?"

62. Murray, *Hamlet on the Holodeck*, 71.

63. Bogost, *Persuasive Games*.

64. If Zillow is in fact indifferent to the truth value of its Zestimate, this would meet the formal definition of *bullshit*. Frankfurt, *On Bullshit*. After all, the company does not profit from being right. Instead, it profits from your repeat visits and the advertising that attention supports. Thanks to information scholar Brian Butler for pointing this out during a conversation at the 2018 iConference.

65. Zestimates are not calculated for a significant amount of low-cost housing. Why? What does this tell us about Zillow's agenda?

66. This definition of digital civics comes from Carl DiSalvo, during a presentation at Georgia Tech, September, 22, 2016. For more on digital approaches to civics, see Gordon and Mihailidis, *Civic Media*.

67. There are several ongoing efforts at Georgia State University to trace these trends. See https://atlmaps.com/; http://digitalcollections.library.gsu.edu/cdm/planningatl; http:/dh2016 .adho.org/abstracts/277.

68. Housing Justice League and Research|Action Cooperative, "BeltLining."

69. See Lands, *Culture of Property*; "From Herndon Homes to two Georgia Domes."

70. Ponczek and Wei, "The 10 Most Unequal Cities in America."

71. The images in this chapter were prototyped early on using the Processing development environment, a platform created by Ben Fry and Casey Raes in order to enable visually oriented people to program graphics using Java—a preexisting object-oriented programming language. They were later finalized using JavaScript in collaboration with Peter Polack, a graduate student at Georgia Tech.

72. To locate Fulton County tax data, see http://qpublic9.qpublic.net/ga_display_dw.php?county= ga_fulton&KEY=14%20004700060091&show_history=1&.

73. See http://www.housingjusticeleague.org.

74. Although the technique dates back to the nineteenth century, the term *small multiple* is best explained in Tufte, *The Visual Display of Quantitative Information*.

75. The Fulton County tax assessor explained in a 2016 interview with the author that such data is error prone.

76. These median values are reported by Trulia, so take them with a grain of salt. For Trulia market trends, see https://www.trulia.com/real_estate/Atlanta-Georgia/market-trends/.

77. Atkinson, "Does Gentrification Help or Harm Urban Neighbourhoods?"

78. See Desmond, *Evicted*.

79. The Department of Housing and Urban Development defines affordable housing as housing that is accessible to those earning the median income. This definition is highly contested, for it means that half of the population can't afford it. This is not the same as the median housing price, which is probably much higher. HUD, "United States Department of Housing and Urban Development.

80. Even the founders of Zillow acknowledge that listings are full of "hidden meanings." Rascoff and Humphries, *Zillow Talk*.

81. I am currently supervising work by Georgia Tech student Eric Corbett to study the coded language about gentrification present in Zillow listings. This work focuses on neighborhoods near the Old Fourth Ward. Corbett's preliminary findings reveal a wide range of variation in how neighborhood change is represented. Corbett has looked at listings from 2012 to 2015. He collected 246 each from Cabbagetown and Reynoldstown. Of all the listings for Cabbagetown, 88 mentioned the neighborhood, and 68 of those characterized the neighborhood in social terms relating to discourses of gentrification. Meanwhile, Reynoldstown listings mentioned the neighborhood 82 times—65 of which portrayed the neighborhood in specific terms. The descriptors used in listings in these neighborhoods, however, were not equivalent. Cabbagetown was described as *historic* in 17 separate listings. The same term showed up only twice for Reynoldstown. Meanwhile, terms such as *new* and *hot* were much more prevalent in Reynoldstown. As of this writing, the full study remains unpublished.

82. See Zukin, Lindeman, and Hurson, "The Omnivore's Neighborhood?"

83. Seehttps://dor.georgia.gov/property-tax-rates.

84. For more on how mass appraisals are conducted, see https://dor.georgia.gov/property-tax-rates.

85. Anything over 30 percent of one's income is considered unaffordable by the Department of Housing and Urban Development. HUD, "United States Department of Housing and Urban Development."

86. See https://dor.georgia.gov/property-tax-exemptions.

87. The rules are different in other cities.

88. Landaw, "Atlanta Declared Renters State of Emergency."

89. Housing Justice League and Research|Action Cooperative, "BeltLining."

90. "The Atlanta BeltLine Project."

91. Housing Justice League and Research|Action Cooperative, "BeltLining," 32.

92. Ibid.

93. For further reading about the Atlanta BeltLine and its unequal impact on Intown neighborhoods, see Immergluck and Balan, "Sustainable for Whom?"

CHAPTER 6

1. See Porter, *Trust in Numbers*; Bowker, *Memory Practices in the Sciences*; Galison, "Limits of Localism.'

2. Simon, *The Sciences of the Artificial*.

3. Schön, *The Reflective Practitioner*, 175.

4. See Sengers et al., "Reflective Design."

5. Agre, *Computation and Human Experience*.

6. Dunne and Raby, *Design Noir*.

7. DiSalvo, *Adversarial Design*. For more on how critical design has been taken up and debated among researchers in human-computer interaction, see Bardzell and Bardzell, "What Is 'Critical' about Critical Design?"; Pierce et al., "Expanding and Refining Design and Criticality in HCI."

8. For a review of the ways in which design can be critical without replicating humanistic approaches based in other media, see Pierce et al., "Expanding and Refining Design and Criticality in HCI."

9. Johanna Drucker provokes us to reconsider our use of the term *data* and offers suggestions for design. Drucker, "Humanities Approaches to Graphical Display." Geoffrey Bowker and Susan Leigh Star's book ends with insightful "design exigencies." Bowker and Star, *Sorting Things Out*, 324. Jeffrey Bardzell and Shaowen Bardzell also offer design implications for those seeking critical reflection. Bardzell and Bardzell, "What Is 'Critical' about Critical Design?" Catherine Ignazio and Lauren Klein supply prescriptions for a critical and feminist approach to data visualization. D'Ignazio and Klein, "Feminist Data Visualization."

10. Dourish, *Where the Action Is*.

11. One danger of offering principles is the "not invented here" problem. Practitioners want to make their own interpretations of theory in their own ways. Moreover, practitioners are

more likely to remember models. As the *Oxford English Dictionary* defines it, "Model (n) A thing used as an example to follow or imitate." For further reading on models in science, see Sismondo, "Models, Simulations, and Their Objects." For further reading on models in education, see Papert, *Mindstorms*. And in design, see Loukissas, *Co-Designers*.

12. For more on how architectural precedents are used in order to convey design ideas, see Rowe, *The Mathematics of the Ideal Villa and Other Essays*; Le Corbusier and Etchells, *Towards a New Architecture*.
13. In recent years, software such as Tableau and software development toolkits such as the JavaScript library D3 have significantly flattened out and generalized the way that data are presented. In many ways, these tools are extraordinary accomplishments and positive contributions, but they should not be accepted uncritically.
14. Norman, *The Design of Everyday Things*, 3.
15. Mendes and Allison, "Bear 71 VR." For a full analysis of this project, see Ray, "Rub Trees, Crittercams, and GIS."
16. Mendes and Allison, "Bear 71 VR."
17. I taught several classes on the subject of data documentaries at Georgia Tech in 2015–2016.
18. Gabrys, *Program Earth*.
19. Thorp, "A Sort of Joy."
20. Ibid.
21. Lupi and Posavec, *Dear Data*, ix.
22. Ibid., x.
23. Rhodes, "This Guy Obsessively Recorded His Private Data for 10 Years."
24. Lupi and Posavec, *Dear Data*, 2.
25. Ibid., 5.
26. Lupi and Posavec, Dear Data.
27. See Laura Kurgan's analysis of Google Earth. Kurgan, *Close Up at a Distance*.
28. Dosemagen, Warren, and Wylie, "Grassroots Mapping." "Civic science" is a reference to Fortun and Fortun, "Scientific Imaginaries and Ethical Plateaus in Contemporary U.S. Toxicology."
29. Dosemagen, Warren, and Wylie, "Grassroots Mapping."
30. Anderson et al., "A Grassroots Remote Sensing Toolkit Using Live Coding, Smartphones, Kites, and Lightweight Drones."
31. For an explanation of how algorithms used to predict recidivism by the criminal justice system can be racist, see Diakopoulos and Friedler, "We Need to Hold Algorithms Accountable."
32. For an example of how counterdata have been used to reveal the biases in facial recognition algorithms, see Buolamwini and Gebru, "Gender Shades." For an example of a counteralgorithm designed to "obfuscate browsing data and protect users from tracking by advertisers," see Nissenbaum, Howe, and Zer-Aviv, "AdNauseam."

33. Brown, "Slave Revolt in Jamaica, 1760–1761," http://revolt.axismaps.com/project.html, accessed August 28, 2018.

34. As Brown explains, "Scholars working in subaltern history rarely have the kind of big databases that inspire projects in text mining, topic modeling, or network analysis." Brown, "Mapping a Slave Revolt," 138.

35. Ibid.

36. Manovich, "Database as Symbolic Form," 85.

37. Brown, "Mapping a Slave Revolt," 138.

38. Anti-Eviction Mapping Project, "About."

39. Ibid.

40. See Loukissas, "Keepers of the Geometry."

41. There is an extensive literature on participatory methods and their potential uses in civic design. For an example, see LeDantec, "Infrastructuring and the Formation of Publics in Participatory Design."

42. Simon, *The Sciences of the Artificial*.

CHAPTER 7

1. Totty, "The Rise of the Smart City"; Jackson, "Columbus under Construction to Become America's First 'Smart City.'" For a more comprehensive look at smart cities see Townsend, *Smart Cities*.

2. Loukissas and Pollock, "After Big Data Failed."

3. University of Texas at Austin, Texas Advanced Computing Center, "The Future of Search Engines."

4. Cadwalladr, "Google, Democracy, and the Truth about Internet Search."

5. See Google's web page "Our Company | Google."

6. Noble, *Algorithms of Oppression*, 14.

7. "What these right wing news sites have done ... is what most commercial websites try to do. They try to find the tricks that will move them up Google's PageRank system. They try and 'game' the algorithm." Cadwalladr, "Google, Democracy, and the Truth about Internet Search."

8. For more on how to tame everyday algorithms for the public good, see Eubanks, *Automating Inequality*; O'Neil, *Weapons of Math Destruction*; Pasquale, *The Black Box Society*.

9. For a primer on adversarial approaches to design, see DiSalvo, *Adversarial Design*.

10. See Bearne, "Is the 'Gig Economy' Turning Us All into Freelancers?"; Armstrong, "The People Making the On-Demand Economy Work."

11. For an overview of open data initiatives, see Kitchin, *The Data Revolution*.

12. I have started to explore this question through the development of *civic data guides* in collaboration with Catherine D'Ignazio, an assistant professor of civic media and data visualization at Emerson College, and Bob Gradeck, who manages the Western Pennsylvania Regional Data Center project at the University of Pittsburgh.

13. See Gradeck, "Data User Guides." Gradeck and his colleagues at the Western Pennsylvania Regional Data Center have already produced more than forty guides at the time of this writing. Their work has been inspired by other guides to civic data, such as Pettit and Droesch, "A Guide to Home Mortgage Disclosure Act Data," and Smart Chicago Collaborative, "Crime and Punishment in Chicago." For parallel work in this area, also see Gebru et al., "Datasheets for Datasets."

14. See data guides offered by the US Department of Education and US Centers for Disease Control and Prevention.

15. See Gradeck, "Data User Guides."

16. For an overview of ethical issues in these areas, see boyd and Crawford, "Critical Questions for Big Data"; Kitchin, *The Data Revolution*.

17. Bowker, *Memory Practices in the Sciences*, 184.

18. Bellacasa, *Matters of Care*, 100.

19. On "instrumental rationality" and "datafication," see Mattern, "Methodolatry and the Art of Measure."

20. Ortner, *Anthropology and Social Theory*, 43.

BIBLIOGRAPHY

Agnew, John. “Representing Space: Space, Scale, and Culture in Social Science.” In *Place/Culture/Representation*, edited by James S. Duncan and David Ley, 251–272. New York: Routledge, 2013.

Agre, Philip E. *Computation and Human Experience*. New York: Cambridge University Press, 1997.

Albright, Jonathan. “Itemized Posts and Historical Engagement: 6 Now-Closed FB Pages.” Accessed December 9, 2017. https://public.tableau.com/views/FB4/TotalReachbyPage?%3Aembed=y&%3AshowVizHome=no&%3Adisplay_count=y&%3Adisplay_static_image=y&%3AbootstrapWhenNotified=true.

Ali, Christopher. *Media Localism: The Policies of Place*. Urbana: University of Illinois Press, 2017.

Ananny, Mike, and Niall Winters. “Designing for Development: Understanding One Laptop per Child in Its Historical Context.” In *Information and Communication Technologies and Development*, 1–12. New York: Institute for Electrical and Electronics Engineers, 2007.

Anderson, Chris. “The End of Theory: The Data Deluge Makes the Scientific Method Obsolete.” WIRED, June 23, 2008. http://www.wired.com/2008/06/pb-theory/.

Anderson, Kyle, Dilys Griffiths, Leon Debell, Simon Hancock, Jeffrey P. Duffy, Jamie D. Shutler, Wolfgang Reinhardt, and Alison L. Griffiths. “A Grassroots Remote Sensing Toolkit Using Live Coding, Smartphones, Kites, and Lightweight Drones.” *PLoS One* 11, no. 5 (May 4, 2016): e0151564.

Anti-Eviction Mapping Project. “About.” Accessed December 7, 2017. https://www.antievictionmap.com/about/.

Armstrong, Stephen. “The People Making the On-Demand Economy Work.” WIRED, December 2015. http://www.wired.co.uk/article/on-demand-economy-airbnb-uber-taskrabbit.

Atkinson, Rowland. “Does Gentrification Help or Harm Urban Neighbourhoods?: An Assessment of the Evidence-Base in the Context of the New Urban Agenda.” ESRC Centre for Neighbourhood Research, June 2002. http://neighbourhoodchange.ca/wp-content/uploads/2011/06/Atkinson-2002-Gentrification-Review.pdf.

“The Atlanta BeltLine Project.” *Atlanta BeltLine* (blog). Accessed May 10, 2018. https://beltline.org/about/the-atlanta-beltline-project/.

Barad, Karen. *Meeting the Universe Halfway: Quantum Physics and the Entanglement of Matter and Meaning*. Durham, NC: Duke University Press, 2007.

Bardzell, Jeffrey, and Shaowen Bardzell. “What Is ‘Critical’ about Critical Design?” In *Proceedings of the SIGCHI Conference on Human Factors in Computing Systems*, 3297–3306. New York: ACM, 2013.

Battles, Matthew. *Library: An Unquiet History*. New York: W. W. Norton and Company, 2003.

Battles, Matthew. *Tree: Object Lessons*. London: Bloomsbury Academic, 2017.

Battles, Matthew, and Yanni Alexander Loukissas. "Data Artefacts: Tracking Knowledge-Ordering Conflicts through Visualization." In *Classification and Visualization: Interfaces to Knowledge*, edited by Aida Slavic, Salah Almila Akdag, and Sylvie Davies, 243–258. Würzburg: Ergon Verlag, 2013.

Baudrillard, Jean. *Simulacra and Simulation*. Translated by Sheila Faria Glaser. Ann Arbor: University of Michigan Press, 1984.

Bearne, Suzanne. "Is the 'Gig Economy' Turning Us All into Freelancers?" BBC News, May 20, 2016. http://www.bbc.com/news/business-36321826.

Bellacasa, Maria Puig de la. "Matters of Care in Technoscience: Assembling Neglected Things." *Social Studies of Science* 41, no. 1 (February 1, 2011): 85–106.

Bellacasa, Maria Puig de la. *Matters of Care: Speculative Ethics in More Than Human Worlds*. Minneapolis: University of Minnesota Press, 2017.

Belsky, Scott. "The Interface Layer: Where Design Commoditizes Tech." *Medium* (blog), May 30, 2014. https://medium.com/positiveslope/the-interface-layer-when-design-commoditizes-tech-e7017872173a.

Benedikt, Michael. *Cyberspace: First Steps*. Cambridge, MA: MIT Press, 1991.

Benkler, Yochai, Robert Faris, and Hal Roberts. "Study: Breitbart-Led Right-Wing Media Ecosystem Altered Broader Media Agenda." *Columbia Journalism Review*, March 3, 2017. https://www.cjr.org/analysis/breitbart-media-trump-harvard-study.php.

Berners-Lee, Tim, Robert Cailliau, Ari Luotonen, Henrik Frystyk Nielsen, and Arthur Secret. "The World-Wide Web." *Communications of the ACM* 37, no. 8 (August 1994): 76–82.

Bijker, Wiebe E. *Of Bicycles, Bakelites, and Bulbs: Toward a Theory of Sociotechnical Change*. Cambridge, MA: MIT Press, 1995.

Binder, Jeffrey M. "Alien Reading: Text Mining, Language Standardization, and the Humanities." In *Debates in the Digital Humanities*, edited by Matthew K. Gold and Lauren F. Klein, 201–217. Minneapolis: University of Minnesota Press, 2016.

Bird, Steven, Ewan Klein, and Edward Loper. "Preface." In *Natural Language Processing with Python*. Accessed June 12, 2017. http://www.nltk.org/book/ch00.html.

Boczkowski, Pablo J. *Digitizing the News: Innovation in Online Newspapers*. Cambridge, MA: MIT Press, 2005.

Boczkowski, Pablo J., and Zizi Papacharissi, eds. *Trump and the Media*. Cambridge, MA: MIT Press, 2018.

Bogost, Ian. *Persuasive Games: The Expressive Power of Videogames*. Cambridge, MA: MIT Press, 2010.

Boguraev, Bran, Ted Briscoe, John Carroll, David Carter, and Claire Grover. "The Derivation of a Grammatically Indexed Lexicon from the Longman Dictionary of Contemporary English." In *Proceedings of the 25th Annual Meeting on Association for Computational Linguistics*, 193–200. Stroudsburg, PA: Association for Computational Linguistics, 1987.

Borges, Jorge Luis. *Collected Fictions*. Translated by Andrew Hurley. New York: Penguin Books, 1999.

Borgman, Christine L. *Big Data, Little Data, No Data: Scholarship in the Networked World*. Cambridge, MA: MIT Press, 2015.

Borkin, Michelle A., Azalea A. Vo, Zoya Bylinskii, Phillip Isola, Shashank Sunkavalli, Aude Oliva, and Hanspeter Pfister. "What Makes a Visualization Memorable?" *IEEE Transactions on Visualization and Computer Graphics* 19, no. 12 (December 2013): 2306–2315.

Bowker, Geoffrey. "Biodiversity Datadiversity." *Social Studies of Science* 30, no. 5 (October 1, 2000): 643–683.

Bowker, Geoffrey. *Memory Practices in the Sciences*. Cambridge, MA: MIT Press, 2006.

Bowker, Geoffrey, and Susan Leigh Star. *Sorting Things Out: Classification and Its Consequences*. Cambridge, MA: MIT Press, 1999.

boyd, danah. *It's Complicated: The Social Lives of Networked Teens*. New Haven, CT: Yale University Press, 2014.

boyd, danah, and Kate Crawford. "Critical Questions for Big Data." *Information Communication and Society* 15, no. 5 (June 1, 2012): 662–679.

Bratton, Benjamin H. *The Stack: On Software and Sovereignty*. Cambridge, MA: MIT Press, 2016.

Brock, André. "From the Blackhand Side: Twitter as a Cultural Conversation." *Journal of Broadcasting and Electronic Media* 56, no. (2012): 529–549.

Brown, Vincent. "Mapping a Slave Revolt: Visualizing Spatial History through the Archives of Slavery." *Social Text* 33, no. 4 (December 1, 2015): 134–141.

Brown, Vincent. "Slave Revolt in Jamaica, 1760–1761: A Cartographic Narrative." Accessed May 15, 2018. http://revolt.axismaps.com/.

Buckland, Michael K. "Information as Thing." *Journal of the American Society for Information Science* 42, no. 5 (June 1, 1991): 351–360.

Buell, Lawrence. *The Future of Environmental Criticism: Environmental Crisis and Literary Imagination*. Hoboken, NJ: John Wiley and Sons, 2009.

Buolamwini, Joy, and Timnit Gebru. "Gender Shades: Intersectional Accuracy Disparities in Commercial Gender Classification." In *Conference on Fairness, Accountability, and Transparency*, 77–91. 2018. http://proceedings.mlr.press/v81/buolamwini18a.html.

Bush, Vannevar. "As We May Think." *Atlantic*, July 1945. http://www.theatlantic.com/magazine/archive/1945/07/as-we-may-think/303881/.

Cadwalladr, Carole. "Google, Democracy, and the Truth about Internet Search." *Guardian*, December 4, 2016, Technology sec. https://www.theguardian.com/technology/2016/dec/04/google-democracy-truth-internet-search-facebook.

Call, Rob, Denechia Powell, and Sarah Heck. "Blackstone: Atlanta's Newest Landlord," April 2014. https://righttothecity.org/blackstone-atlantas-newest-landlord/.

Castells, Manuel. *The Information Age: Economy, Society, and Culture.* Vol. 1 of *The Rise of the Network Society*. 2nd ed. Chichester, UK: Wiley-Blackwell, 2009.

Cetina, Karin Knorr. *Epistemic Cultures: How the Sciences Make Knowledge*. Cambridge, MA: Harvard University Press, 1999.

Chalabi, Mona. "'Data Is' vs. 'Data Are.'" FiveThirtyEight, March 17, 2014. https://fivethirtyeight.com/features/data-is-vs-data-are/.

Chan, Anita Say. *Networking Peripheries: Technological Futures and the Myth of Digital Universalism*. Cambridge, MA: MIT Press, 2016.

Confessore, Nicholas, and Daisuke Wakabayashi. "How Russia Harvested American Rage to Reshape U.S. Politics." *New York Times*, October 9, 2017, Technology sec. https://www.nytimes.com/2017/10/09/technology/russia-election-facebook-ads-rage.html.

Cosgrove, Denis. *Geography and Vision: Seeing, Imagining, and Representing the World.* London: I. B. Tauris, 2008.

Costanza-Chock, Sasha, and Pablo Rey-Mazón. "PageOneX: New Approaches to Newspaper Front Page Analysis." *International Journal of Communication* 10, no. 0 (April 28, 2016): 2318–2345.

Cotter, Holland. "The Met Reimagines the American Story." *New York Times*, January 15,2012.http://www.nytimes.com/2012/01/16/arts/design/metropolitan-museum-of-arts-new-american-wing-galleries-review.html.

Crampton, Jeremy W., and John Krygier. "An Introduction to Critical Cartography." *ACME: An International E-Journal for Critical Geographies* 4 (July 16, 2010).

Crampton, Jeremy W. *Mapping: A Critical Introduction to Cartography and GIS*. Hoboken, NJ: John Wiley and Sons, 2011.

Crawford, Kate. "The Anxieties of Big Data." *New Inquiry* (blog), May 30, 2014. https://thenewinquiry.com/the-anxieties-of-big-data/.

Cresswell, Tim. *In Place/Out of Place*. Minneapolis: University of Minnesota Press, 1996.

Culler, Jonathan. "The Closeness of Close Reading." *ADE Bulletin* 149 (January 1, 2010): 20–25.

Dalton, Craig, and Jim Thatcher. "What Does a Critical Data Studies Look Like, and Why Do We Care?" *Society and Space—Environment and Planning D* (blog), 2014. http://societyandspace.org/2014/05/12/what-does-a-critical-data-studies-look-like-and-why-do-we-care-craig-dalton-and-jim-thatcher/.

Darton, Robert. "Jefferson's Taper: A National Digital Library." *New York Review of Books*, November 24, 2011. http://www.nybooks.com/articles/2011/11/24/jeffersons-taper-national-digital-library/.

Del Tredici, Peter. *Wild Urban Plants of the Northeast: A Field Guide*. Ithaca, NY: Cornell University Press, 2010.

Del Tredici, Peter, and Alice Kitajima. "Finding a Replacement for the Eastern Hemlock: Research at the Arnold Arboretum." *Arnoldia,* no. 63 (2) (2004): 33–39.

Desmond, Matthew. *Evicted.* New York: Crown, 2016.

Dewey, John. *The Public and Its Problems*. New York: Henry Holt and Company, 1927.

Dey, Anind K. "Understanding and Using Context." *Personal and Ubiquitous Computing* 5, no. 1 (January 2001): 4–7.

Diakopoulos, Nicholas. "Algorithmic Accountability: On the Investigation of Black Boxes." Tow Center for Digital Journalism, December 3, 2014. https://towcenter.org/research/algorithmic-accountability-on-the-investigation-of-black-boxes-2/.

Diakopoulos, Nicholas, and Sorelle Friedler. "We Need to Hold Algorithms Accountable—Here's How to Do It." *MIT Technology Review*, November 17, 2016. https://www.technologyreview.com/s/602933/how-to-hold-algorithms-accountable/.

"Digital Public Library of America Strategic Plan, 2015 through 2017." January 2015. https://pro.dp.la/about-dpla-pro/strategic-plan.

D'Ignazio, Catherine, and Lauren F. Klein. "Feminist Data Visualization." IEEE InfoViz Workshop, 2017.

DiSalvo, Carl. *Adversarial Design*. Cambridge, MA: MIT Press, 2012.

Dosemagen, Shannon, Jeffrey Warren, and Sara Wylie. "Grassroots Mapping: Creating a Participatory Map-Making Process Centered on Discourse." *Public Laboratory* (blog). Accessed December 7, 2017. http://www.joaap.org/issue8/GrassrootsMapping.htm.

Douglas, Mary. *Purity and Danger: An Analysis of the Concepts of Pollution and Taboo*. New York: Routledge and Kegan Pau, 1978.

Dourish, Paul. "Re-Space-Ing Place: 'Place' and 'Space' Ten Years On." In *Proceedings of the 2006 20th Anniversary Conference on Computer Supported Cooperative Work*, 299–308. New York: ACM, 2006.

Dourish, Paul. *The Stuff of Bits*. Cambridge, MA: MIT Press, 2017.

Dourish, Paul. "What We Talk about When We Talk about Context." *Personal and Ubiquitous Computing* 8, no. 1 (February 2004): 19–30.

Dourish, Paul. *Where the Action Is*. Cambridge, MA: MIT Press, 2001.

Downey, Gregory J. "Making Media Work: Time, Space, Identity, and Labor in the Analysis of Information and Communication Infrastructures." In *Media Technologies: Essays on Communication, Materiality, and Society*, edited by Tarleton Gillespie, Pablo J. Boczkowski, and Kirsten A. Foot, 141–166. Cambridge, MA: MIT Press, 2014.

"DPLA Metadata Application Profile." Accessed December 18, 2017. https://dp.la/info/developers/map/.

Drucker, Johanna. *Graphesis: Visual Forms of Knowledge Production*. Cambridge, MA: Harvard University Press, 2014.

Drucker, Johanna. "Humanities Approaches to Graphical Display." *Digital Humanities Quarterly* 5, no. 1 (2011). http://www.digitalhumanities.org/dhq/vol/5/1/000091/000091.html.

"Dublin Core Metadata Initiative." Accessed December 18, 2017. http://www.dublincore.org/.

Dunne, Anthony. *Hertzian Tales: Electronic Products, Aesthetic Experience, and Critical Design*. Cambridge, MA: MIT Press, 2008.

Dunne, Anthony, and Fiona Raby. *Design Noir: The Secret Life of Electronic Objects*. London: Birkhäuser, 2001.

Dunne, Anthony, and Fiona Raby. *Speculative Everything: Design, Fiction, and Social Dreaming*. Cambridge, MA: MIT Press, 2013.

Edwards, Paul N. *A Vast Machine: Computer Models, Climate Data, and the Politics of Global Warming*. Cambridge, MA: MIT Press, 2010.

Edwards, Paul N., Matthew S. Mayernik, Archer L. Batcheller, Geoffrey C. Bowker, and Christine L. Borgman. "Science Friction: Data, Metadata, and Collaboration." *Social Studies of Science* 41, no. 5 (2011): 667–690.

Eisenstein, Jacob. "What to Do about Bad Language on the Internet." In *Proceedings of the 2013 Conference of the North American Chapter of the Association for Computational Linguistics: Human Language Technologies*, 359–369. Stroudsburg, PA: Association for Computational Linguistics, 2013.

Elmborg, James. "Critical Information Literacy: Implications for Instructional Practice." *Journal of Academic Librarianship* 32, no. 2 (March 2006): 192–199.

Eubanks, Virginia. *Automating Inequality: How High-Tech Tools Profile, Police, and Punish the Poor*. New York: St. Martin's Press, 2018.

Fake News Challenge. Accessed December 9, 2017. http://www.fakenewschallenge.org/.

Fiegerman, Seth, and Dylan Byers. "Facebook, Twitter, Google Testify before Congress."CNNMoney, October 31, 2017. http://money.cnn.com/2017/10/31/media/facebook-twitter-google-congress/index.html.

Fillmore, Charles. "Encounters with Language." *Computational Linguistics* 38, no. 4 (December 2012): 701–718.

Finn, Ed. *What Algorithms Want: Imagination in the Age of Computing*. Cambridge, MA: MIT Press, 2018.

Fortun, Kim, and Michael Fortun. "Scientific Imaginaries and Ethical Plateaus in Contemporary U.S. Toxicology." *American Anthropologist* 107, no. 1 (January 7, 2008): 43–54.

Foucault, Michel. *The Archaeology of Knowledge: And the Discourse on Language*. New York: Pantheon Books, 1972.

Frampton, Kenneth. "Towards a Critical Regionalism: Six Points for an Architecture of Resistance." In *Postmodern Culture*, edited by Hal Foster. 16–30. London: Pluto, 2001.

Frankfurt, Harry G. *On Bullshit*. Princeton, NJ: Princeton University Press, 2005.

"From Herndon Homes to Two Georgia Domes: 100 Years of Westside History." Georgia Tech, Ivan Allen College of Liberal Arts. Accessed July 12, 2018. http://www.iac.gatech.edu/news-events/events/2015/11/herndon-homes-georgia-domes-100-years-westside-history/472161.

Fry, Ben. *Visualizing Data: Exploring and Explaining Data with the Processing Environment*. Sebastopol, CA: O'Reilly Media, 2008.

Gabrys, Jennifer. *Program Earth: Environmental Sensing Technology and the Making of a Computational Planet*. Minneapolis: University of Minnesota Press, 2016.

Gabrys, Jennifer, Helen Pritchard, and Benjamin Barratt. "Just Good Enough Data: Figuring Data Citizenships through Air Pollution Sensing and Data Stories." *Big Data and Society* 3, no. 2 (December 1, 2016). https://doi.org/10.1177/2053951716679677.

Galison, Peter. "Limits of Localism: The Scale of Sight." In *What Reason Promises: Essays on Reason, Nature, and History*, edited by Wendy Doniger, Peter Galison, and Susan Neiman, 155–170. Berlin: De Gruyter, 2016.

Galison, Peter, and Emily Thompson, eds. *The Architecture of Science*. Cambridge, MA: MIT Press, 1999.

Galloway, Alexander R. *The Interface Effect*. Cambridge, UK: Polity, 2012.

Gebru, Timnit, Jamie Morgenstern, Briana Vecchione, Jennifer Wortman Vaughan, Hanna Wallach, Hal Daumeé III, and Kate Crawford. "Datasheets for Datasets." March 23, 2018. http://arxiv.org/abs/1803.09010.

Geertz, Clifford. *Available Light: Anthropological Reflections on Philosophical Topics*. Princeton, NJ: Princeton University Press, 2012.

Geertz, Clifford. *The Interpretation of Cultures: Selected Essays*. New York: Basic Books, 1973.

Geertz, Clifford. "Local Knowledge and Its Limits." *Yale Journal of Criticism* 5, no. 2 (1992): 129–135.

Geertz, Clifford. *Local Knowledge: Further Essays in Interpretive Anthropology*. New York: Basic Books, 2008.

Gibson, William. *Neuromancer*. New York: Ace, 1984.

Gieryn, Thomas F. "A Space for Place in Sociology." *Annual Review of Sociology* 26, no. 1 (August 2000): 463–496.

Gillespie, Tarleton. "The Relevance of Algorithms." In *Media Technologies: Essays on Communication, Materiality, and Society*, edited by Tarleton Gillespie, Pablo J. Boczkowski, and Kirsten A. Foot, 167–194 Cambridge, MA: MIT Press, 2014.

Gillespie, Tarleton, Pablo J. Boczkowski, and Kirsten A. Foot, eds. *Media Technologies: Essays on Communication, Materiality, and Society*. Cambridge, MA: MIT Press, 2014.

Gitelman, Lisa, ed. *"Raw Data" Is an Oxymoron*. Cambridge, MA: MIT Press, 2013.

Gnoli, C. "Metadata about What? Distinguishing between Ontic, Epistemic, and Documental Dimensions in Knowledge Organizations." *Knowledge Organization* 39, no. 4 (January 1, 2012): 268–275.

Goffman, Erving. *The Presentation of Self in Everyday Life*. New York: Anchor, 1959.

Gordon, Eric, and Paul Mihailidis, eds. *Civic Media: Technology, Design, Practice*. Cambridge, MA: MIT Press, 2016.

Gradeck, Bob. "Data User Guides." Western Pennsylvania Regional Data Center. Accessed December 7, 2017. https://www.wprdc.org/data-user-guides/.

Graeber, David. *Toward an Anthropological Theory of Value: The False Coin of Our Own Dreams*. New York: Palgrave, 2016.

Graham, Stephen. "The End of Geography or the Explosion of Place? Conceptualizing Space, Place, and Information Technology." *Progress in Human Geography* 22, no. 2 (April 1, 1998): 165–185.

Gusterson, Hugh. *Nuclear Rites: A Weapons Laboratory at the End of the Cold War*. Berkeley: University of California Press, 1998.

Hacking, Ian. "A Tradition of Natural Kinds." *Philosophical Studies: An International Journal for Philosophy in the Analytic Tradition* 61, no. 1–2 (1991): 109–126.

Haffner, Jeanne. *The View from Above: The Science of Social Space*. Cambridge, MA: MIT Press, 2013.

Hall, Peter. "Critical Visualization." In *Design and the Elastic Mind*, edited by Paola Antonelli, 120–131. New York: Museum of Modern Art, 2008.

Halpern, Orit. *Beautiful Data: A History of Vision and Reason since 1945*. Durham, NC: Duke University Press, 2015.

Hanan, Joshua S. "Home Is Where the Capital Is: The Culture of Real Estate in an Era of Control Societies." *Communication and Critical/Cultural Studies* 7, no. 2 (June 1, 2010): 176–201.

Hancox, Peter. "A Brief History of NLP." Accessed December 6, 2017. http://www.cs.bham.ac.uk/~pjh/sem1a5/pt1/pt1_history.html.

Haraway, Donna. "Situated Knowledges: The Science Question in Feminism and the Privilege of Partial Perspective." *Feminist Studies* 14, no. 3 (1988): 575–599.

Harding, Sandra G. "Is Science Multicultural?: Challenges, Resources, Opportunities, Uncertainties." *Configurations* 2, no. 2 (May 1, 1994): 301–330.

Hayles, N. Katherine. *How We Became Posthuman: Virtual Bodies in Cybernetics, Literature, and Informatics*. Chicago: University of Chicago Press, 2008.

Hookway, Branden. *Interface*. Cambridge, MA: MIT Press, 2014.

Housing Justice League, and Research|Action Cooperative. "BeltLining: Gentrification, Broken Promises, and Hope on Atlanta's Southside." October 2017.

Houston, Keith. "The Mysterious Origins of Punctuation." BBC News, September 2, 2015. http://www.bbc.com/culture/story/20150902-the-mysterious-origins-of-punctuation.

Houston, Laura, and Steven Jackson. "Caring for the Next Billion Mobile Handsets: Opening Proprietary Closures through the Work of Repair." In *Proceedings of ACM Information and Communication Technologies for Development*. Ann Arbor: ACM, 2016.

Hovy, Dirk, and Shannon L. Spruit. "The Social Impact of Natural Language Processing." In *Annual Meeting of the Association for Computational Linguistics: Proceedings of the Conference*, 591–598. Berlin: ACL, 2016.

Howard, Jennifer. "Book Lovers Record Traces of 19th-Century Readers." Blog. *Chronicle of Higher Education*, May 5, 2014. https://www.chronicle.com/blogs/wiredcampus/book-lovers-record-traces-of-19th-century-readers/52415.

Howell, David. "How Accurate Is Zillow's Zestimate? Not Very, Says One Washington-Area Agent." *Washington Post*, June 10, 2014. https://www.washingtonpost.com/news/where-we-live/wp/2014/06/10/how-accurate-is-zillows-zestimate-not-very-says-one-washington-area-agent.

HUD. "United States Department of Housing and Urban Development (HUD)." Accessed December 7, 2017. https://www.hud.gov/program_offices/comm_planning/affordablehousing.

Humphries, Stan. "How Accurate Is the Zestimate? Zillow Says the Tool Is Helpful When Used the Right Way." *Washington Post*, March 10, 2014. https://www.washingtonpost.com/news/where-we-live/wp/2014/06/10/how-accurate-is-the-zestimate-zillow-says-the-tool-is-helpful-when-used-the-right-way.

Immergluck, Dan, and Tharunya Balan. "Sustainable for Whom? Green Urban Development, Environmental Gentrification, and the Atlanta Beltline." *Urban Geography* 39, no. 4 (April 21, 2018): 546–562.

Immergluck, Dan, Ann Carpenter, and and Abram Lueders. "Declines in Low-Cost Rented Housing Units in Eight Large Southeastern Cities." Federal Reserve Bank of Atlanta, Community and Economic Development Department, 2016.

Irani, Lilly, Janet Vertesi, Paul Dourish, Kavita Philip, and Rebecca E. Grinter. "Postcolonial Computing: A Lens on Design and Development." In *Proceedings of the SIGCHI Conference on Human Factors in Computing Systems*, 1311–1320. New York: ACM, 2010.

Jackson, Katie. "Columbus under Construction to Become America's First 'Smart City.'" Fox News, July 10, 2016. http://www.foxnews.com/tech/2017/07/10/columbus-under-construction-to-become-americas-first-smart-city.html.

Jaenicke, Stefan, and Greta Franzini. "On Close and Distant Reading in Digital Humanities: A Survey and Future Challenges. A State-of-the-Art (STAR) Report." 2015.

Jones, Karen Spärck. "Natural Language Processing: A Historical Review." In *Current Issues in Computational Linguistics: In Honour of Don Walker*, 3–16. Dordrecht: Springer, 1994.

Kalay, Yehuda E., and John Marx. "The Role of Place in Cyberspace." In *Proceedings of the Seventh International Conference on Virtual Systems and Multimedia (VSMM'01)*, 770–779. Washington: IEEE Computer Society, 2001.

Karadkar, Unmil P., Audrey Altman, Mark Breedlove, and Mark Matienzo. "Introduction to the Digital Public Library of America API." In *2016 IEEE/ACM Joint Conference on Digital Libraries*, 285–286. Toronto: IEEE, 2016.

Kay, Alan, and Adele Goldberg. "Personal Dynamic Media." *Computer* 10, no. 3 (1977): 31–47.

Keller, Evelyn Fox. *Making Sense of Life: Explaining Biological Development with Models, Metaphors, and Machines*. Cambridge, MA: Harvard University Press, 2003.

Kirk, Andy. *Data Visualization: A Handbook for Data Driven Design*. Los Angeles: SAGE Publications, 2016.

Kirschenbaum, Matthew G. *Mechanisms: New Media and the Forensic Imagination*. Cambridge, MA: MIT Press, 2012.

Kitchin, Rob. "Big Data, New Epistemologies, and Paradigm Shifts." *Big Data and Society* 1, no. 1 (April 1, 2014): 1–12.

Kitchin, Rob. *The Data Revolution: Big Data, Open Data, Data Infrastructures, and Their Consequences*. Thousand Oaks, CA: SAGE Publications, 2014.

Kitchin, Rob, and Martin Dodge. *Code/Space: Software and Everyday Life*. Cambridge, MA: MIT Press, 2011.

Kitchin, Rob, Justin Gleeson, and Martin Dodge. "Unfolding Mapping Practices: A New Epistemology for Cartography." *Transactions of the Institute of British Geographers* 38, no. 3 (July 1, 2013): 480–496.

Kitchin, Rob, and Tracey P. Lauriault. "Towards Critical Data Studies: Charting and Unpacking Data Assemblages and Their Work." SSRN Scholarly Paper. Rochester, NY: Social Science Research Network, July 30, 2014. http://papers.ssrn.com/abstract=2474112.

Kitchin, Rob, and Gavin McArdle. "What Makes Big Data, Big Data? Exploring the Ontological Characteristics of 26 Datasets." *Big Data and Society* 3, no. 1 (February 17, 2016): 1–10.

Klein, Lauren F. "The Image of Absence: Archival Silence, Data Visualization, and James Hemings." *American Literature* 85, no. 4 (December 1, 2013): 661–688.

Kleinman, Zoe. "Artificial Intelligence: How to Avoid Racist Algorithms." BBC News, April 14, 2017, Technology sec. http://www.bbc.com/news/technology-39533308.

Kovach, Bill, and Tom Rosenstiel. *The Elements of Journalism: What Newspeople Should Know and the Public Should Expect, Completely Updated and Revised*. New York: Three Rivers Press, 2007.

Kurgan, Laura. *Close Up at a Distance: Mapping, Technology, and Politics*. New York: Zone Books, 2013.

Landaw, Laura. "Atlanta Declared Renters State of Emergency." July 13, 2016. http://www.housingjusticeleague.org/atlanta_declared_a_renter_s_state_of_emergency.

Lands, LeeAnn. *Culture of Property: Race, Class, and Housing Landscapes in Atlanta, 1880–1950*. Athens: University of Georgia Press, 2009.

Latour, Bruno. "Drawing Things Together." In *Representation in Scientific Practice*, edited by Michael Lynch and Stephen Woolgar, 19–68. Cambridge, MA: MIT Press, 1990.

Latour, Bruno. *Reassembling the Social: An Introduction to Actor-Network-Theory*. Oxford: Oxford University Press, 2008.

Latour, Bruno. *Science in Action: How to Follow Scientists and Engineers through Society*. Cambridge, MA: Harvard University Press, 1987.

Latour, Bruno, and Steve Woolgar. *Laboratory Life: The Social Construction of Scientific Facts*. Beverly Hills, CA: SAGE Publications, 1979.

Law, John, and Ruth Benschop. “Resisting Pictures: Representation, Distribution, and Ontological Politics.” *Sociological Review* 45, no. S1 (May 1, 1997): 158–182.

Law, John, and Annemarie Mol. “Situating Technoscience: An Inquiry into Spatialities.” *Environment and Planning D: Society and Space* 19, no. 5 (October 1, 2001): 609–621.

Le Corbusier, and Frederick Etchells. *Towards a New Architecture*. Eastford, CT: Martino Publishing, 2014.

LeDantec, Christopher. “Infrastructuring and the Formation of Publics in Participatory Design.” *Social Studies of Science* 43, no. 2 (2013): 241–264.

LeFurgy, Bill. “Digging into a Slice of Digital History.” *The Signal* (blog), January 11, 2012. https://blogs.loc.gov/thesignal/2012/01/digging-into-a-slice-of-digital-history/.

Leurs, Koen. “Feminist and Postcolonial Data Analysis. Using Digital Methods for Ethical, Reflexive, and Situated Socio-Cultural Research.” *International Politics Review* 115, no. 1 (March 1, 2017): 130–154.

Lewis, Britt. “ASU, MIT to Develop New Approaches to Library Print Collections.” *Exploring Value of Print in the Digital Age | ASU Now: Access, Excellence, Impact* (blog), March 14, 2017. https://asunow.asu.edu/20170314-solutions-exploring-value-print-digital-age.

Lewis, Michael. *The Big Short: Inside the Doomsday Machine*. London: Allen Lane, 2010.

Lievrouw, Leah A. “Materiality and Media in Communication and Technology Studies.” In *Media Technologies: Essays on Communication, Materiality, and Society*, edited by Tarleton Gillespie, Pablo J. Boczkowski, and Kirsten A. Foot, 21–53. Cambridge, MA: MIT Press, 2014.

Livingstone, David. *Putting Science in Its Place*. Chicago: University of Chicago Press, 2003.

Lohr, Steve. “The Age of Big Data.” *New York Times*, February 11, 2012, Sunday Review sec. http://www.nytimes.com/2012/02/12/sunday-review/big-datas-impact-in-the-world.html?_r=1.

Loukissas, Yanni Alexander. “All the Homes: Zillow and the Operational Context of Data.” In *Transforming Digital Worlds: iConference 2018*, edited by Gobinda Chowhury, Julie McLeod, Val Gillet, and Peter Willett, 272–281. Cham, Switzerland: Springer, 2018.

Loukissas, Yanni Alexander. *Co-Designers: Cultures of Computer Simulation in Architecture*. London: Routledge, 2012.

Loukissas, Yanni Alexander. “Keepers of the Geometry.” In *Simulation and Its Discontents*, edited by Sherry Turkle, 153–170. Cambridge, MA: MIT Press, 2009.

Loukissas, Yanni Alexander. “A Place for Big Data: Close and Distant Readings of Accessions Data from the Arnold Arboretum.” *Big Data and Society* 3, no. 2 (December 1, 2016). https://doi.org/10.1177/2053951716661365.

Loukissas, Yanni Alexander. “Taking Big Data Apart: Local Readings of Composite Media Collections.” *Information Communication and Society* 20, no. 5 (May 4, 2017): 651–664.

Loukissas, Yanni Alexander, Jason Maron, Marco Zenati, and David Mindell. “Redesigning Postoperative Review.” In *IEEE Healthcare Technology Conference Proceedings*. Houston: IEEE, 2012.

Loukissas, Yanni Alexander, and David Mindell. "Visual Apollo: A Graphical Exploration of Computer-Human Relationships." *Design Issues* 30, no. 2 (April 1, 2014): 4–16.

Loukissas, Yanni Alexander, and Anne Pollock. "After Big Data Failed: The Enduring Allure of Numbers in the Wake of the 2016 US Election." *Engaging Science, Technology, and Society* 3, no. 0 (February 17, 2017): 16–20.

Lupi, Giorgia, and Stefanie Posavec. *Dear Data*. Princeton, NJ: Princeton Architectural Press, 2016.

Lupi, Giorgia, and Stefanie Posavec. Dear Data. Accessed May 17, 2018. http://www.dear-data.com.

Lynch, Michael. *Art and Artifact in Laboratory Science: A Study of Shop Work and Shop Talk in a Research Laboratory*. London: Routledge and Kegan Paul, 1985.

Manovich, Lev. "Database as Symbolic Form." *Convergence* 5, no. 2 (June 1, 1999): 80–99.

Manovich, Lev. *The Language of New Media*. Cambridge, MA: MIT Press, 2002.

Marcus, Mitchell P., Mary Ann Marcinkiewicz, and Beatrice Santorini. "Building a Large Annotated Corpus of English: The Penn Treebank." *Computational Linguistics* 19, no. 2 (June 1993): 313–330.

Marwick, Alice E, and danah boyd. "Networked Privacy: How Teenagers Negotiate Context in Social Media." *New Media and Society* 16, no. 7 (November 1, 2014): 1051–1067. https://doi.org/10.1177/1461444814543995.

Matienzo, Mark, and Amy Rudersdorf. "The Digital Public Library of America Ingestion Ecosystem: Lessons Learned After One Year of Large-Scale Collaborative Metadata Aggregation," 12–23. 2014.

Mattern, Shannon. "Methodolatry and the Art of Measure." *Places Journal*, November 5, 2013. https://doi.org/10.22269/131105.

Mayernik, Matthew S., Archer L. Batcheller, and Christine L. Borgman. "How Institutional Factors Influence the Creation of Scientific Metadata." In *Proceedings of the 2011 iConference*, 417–425. New York: ACM, 2011.

Mayer-Schoenberger, Viktor, and Kenneth Neil Cukier. "The Rise of Big Data." *Foreign Affairs*, September 15, 2015. https://www.foreignaffairs.com/articles/2013-04-03/rise-big-data.

McLuhan, Marshall. *Understanding Media*. London: Sphere, 1967.

Mendes, Jeremy, and Leanne Allison. "Bear 71 VR." Accessed December 7, 2017. https://bear71vr.nfb.ca/.

Mindell, David. *Digital Apollo: Human and Machine in Spaceflight*. Cambridge, MA: MIT Press, 2008.

Mitchell, William J. *City of Bits: Space, Place, and the Infobahn*. Cambridge, MA: MIT Press, 1995.

Mitchell, William J. Me++: *The Cyborg Self and the Networked City*. Cambridge, MA: MIT Press, 2004.

Mody, Cyrus C. M. "A Little Dirt Never Hurt Anyone: Knowledge-Making and Contamination in Materials Science." *Social Studies of Science* 31, no. 1 (February 1, 2001): 7–36.

Mol, Annemarie. *The Body Multiple: Ontology in Medical Practice*. Durham, NC: Duke University Press, 2003.

Morozov, Evgeny. "The Rise of Data and the Death of Politics." *Guardian*, July 19, 2014, Technology sec. https://www.theguardian.com/technology/2014/jul/20/rise-of-data-death-of-politics-evgeny-morozov-algorithmic-regulation.

Muller, Jerry. *The Tyranny of Metrics*. Princeton, NJ: Princeton University Press, 2018.

Murray, Janet H. *Hamlet on the Holodeck*. Cambridge, MA: MIT Press, 1998.

Murray, Janet H., Sergio Goldenberg, Kartik Agarwal, Abraham Doris-Down, Pradyut Pokuri, Nachiketas Ramanujam, and Tarun Chakravorty. "StoryLines: An Approach to Navigating Multisequential News and Entertainment in a Multiscreen Framework." In *Proceedings of the 8th International Conference on Advances in Computer Entertainment Technology*, 92:1–92:2. New York: ACM, 2011.

Nagel, Thomas. *The View from Nowhere*. New York: Oxford University Press, 1986.

Nasukawa, Tetsuya, and Jeonghee Yi. "Sentiment Analysis: Capturing Favorability Using Natural Language Processing." In *Proceedings of the 2nd International Conference on Knowledge Capture*, 70–77. New York: ACM, 2003.

"Natural Language Toolkit—NLTK 3.2.5 Documentation." Accessed December 6, 2017. http://www.nltk.org/.

Negroponte, Nicholas. *Being Digital*. New York: Knopf, 1995.

Negroponte, Nicholas. "One Laptop per Child." *Economist*, November 18, 2005. https://www.economist.com/node/5134619.

Nelson, Theodore H. "Complex Information Processing: A File Structure for the Complex, the Changing, and the Indeterminate." In *Proceedings of the 1965 20th National Conference*, 84–100. New York: ACM, 1965.

Nelson, Theodore H., and Stewart Brand. *Computer Lib/Dream Machines*. Rev. ed. Redmond, CA: Tempus Books, 1987.

Nextdoor. Accessed December 18, 2017. https://nextdoor.com/.

Ng, Andrew. "What Is Machine Learning?" Coursera. Accessed December 10, 2017. https://www.coursera.org/learn/machine-learning/lecture/Ujm7v/what-is-machine-learning.

Nissenbaum, Helen, Daniel C. Howe, and Mushon Zer-Aviv. "AdNauseam—Clicking Ads So You Don't Have To." Accessed April 29, 2018. http://adnauseam.io.

Noble, Safiya. *Algorithms of Oppression: How Search Engines Reinforce Racism*. New York: NYU Press, 2018.

Norman, Donald A. *The Design of Everyday Things*. New York: Basic Books, 2002.

Nye, David E. *American Technological Sublime*. Cambridge, MA: MIT Press, 1996.

Office of the Director of National Intelligence. "Joint DHS, ODNI, FBI Statement on Russian Malicious Cyber Activity." December 29, 2016. https://www.dni.gov/index.php/newsroom/press

-releasespress-releases-2016item/1616-joint-dhs-odni-fbi-statement-on-russian-malicious-cyber-activity?highlight=WyJydXNzaWFuliwicnVzc2lhbnMiXQ==.

O'Neil, Cathy. *Weapons of Math Destruction: How Big Data Increases Inequality and Threatens Democracy*. New York: Crown, 2016.

O'Neil, Kathy, and Rachel Schutt. *Doing Data Science: Straight Talk from the Frontline*. Sebastopol, CA: O'Reilly Media, 2013.

Ortner, Sherry B. *Anthropology and Social Theory: Culture, Power, and the Acting Subject*. Durham, NC: Duke University Press, 2006.

Palfrey, John, and Urs Gasser. *Interop: The Promise and Perils of Highly Interconnected Systems*. New York: Basic Books, 2012.

Pang, Bo, Lillian Lee, and Shivakumar Vaithyanathan. "Thumbs Up?: Sentiment Classification Using Machine Learning Techniques." In *Proceedings of the ACL-02 Conference on Empirical Methods in Natural Language Processing—Volume 10*, 79–86. Stroudsburg, PA: Association for Computational Linguistics, 2002.

Papert, Seymour. *Mindstorms: Children, Computers, and Powerful Ideas*. New York: Basic Books, 1980.

Pariser, Eli. *The Filter Bubble: How the New Personalized Web Is Changing What We Read and How We Think*. New York: Penguin Books, 2012.

Parsons, Talcott. *The Social System*. Florence, MA: Free Press, 1951.

Pasquale, Frank. *The Black Box Society: The Secret Algorithms That Control Money and Information*. Cambridge, MA: Harvard University Press, 2015.

Passi, Samir, and Steven Jackson. "Data Vision: Learning to See through Algorithmic Abstraction." In *Proceedings of the 2017 ACM Conference on Computer Supported Cooperative Work and Social Computing*, 2436–2457. New York: ACM, 2017.

Pettit, Kathryn L. S., and Audrey Droesch. "A Guide to Home Mortgage Disclosure Act Data." Urban Institute, June 4, 2016. https://www.urban.org/research/publication/guide-home-mortgage-disclosure-act-data.

Piaget, Jean. *The Psychology of Intelligence*. New York: Routledge, 1950.

Pierce, James, Phoebe Sengers, Tad Hirsch, Tom Jenkins, William Gaver, and Carl DiSalvo. "Expanding and Refining Design and Criticality in HCI." In *Proceedings of the 33rd Annual ACM Conference on Human Factors in Computing Systems*, 2083–2092. New York: ACM, 2015.

Ponczek, Sarah, and Lu Wei. "The 10 Most Unequal Cities in America." Accessed December 7, 2017. https://www.bloomberg.com/news/articles/2016-10-05/miami-is-the-newly-crowned-most-unequal-city-in-the-u-s.

Porter, Theodore M. *Trust in Numbers: The Pursuit of Objectivity in Science and Public Life*. Princeton, NJ: Princeton University Press, 2001.

Posner, Miriam, and Lauren F. Klein. "Data as Media." *Feminist Media Histories* 3, no. 3 (July 1, 2017): 1–8.

"Public Lab: A DIY Environmental Science Community." Accessed December 5, 2017. https://publiclab.org/.

Ralph, Grishman. "Information Extraction: Capabilities and Challenges. Notes Prepared for the 2012 International Winter School in Language and Speech Technologies, Rovira I Virgili University, Tarragona, Spain," January 21, 2012. https://cs.nyu.edu/grishman/tarragona.pdf.

Ramirez, Edith, Julie Brill, Maureen K. Ohlhausen, Joshua D. Wright, and Terrell McSweeny. "Data Brokers: A Call for Transparency and Accountability." US Federal Trade Commission, May 2014.

Rascoff, Spencer, and and Stan Humphries. *Zillow Talk: Rewriting the Rules of Real Estate*. New York: Grand Central Publishing, 2015.

Ratto, Matt. "Critical Making: Conceptual and Material Studies in Technology and Social Life." *Information Society* 27, no. 4 (July 2011): 252–260.

Ray, Sarah Jaquette. "Rub Trees, Crittercams, and GIS: The Wired Wilderness of Leanne Allison and Jeremy Mendes' Bear 71." *Green Letters* 18, no. 3 (September 2, 2014): 236–253.

Rhodes, Margaret. "This Guy Obsessively Recorded His Private Data for 10 Years." WIRED, October 19, 2015. https://www.wired.com/2015/10/nicholas-felton-obsessively-recorded-his-private-data-for-10-years/.

Ribes, David. "The Rub and Chafe of Maintenance and Repair." *Continent* 6, no. 1 (March 23, 2017): 71–76.

Ribes, David, and Thomas A. Finholt. "The Long Now of Technology Infrastructure: Articulating Tensions in Development," *Journal of the Association for Information Systems* 10, no. 5 (2009): 375–398.

Richtel, Matt. "How Big Data Is Playing Recruiter for Specialized Workers." *New York Times*, April 27, 2013, Technology sec. https://www.nytimes.com/2013/04/28/technology/how-big-data-is-playing-recruiter-for-specialized-workers.html.

"Rise of the Corporate Landlord." Accessed May 9, 2018. https://righttothecity.org/cause/rise-of-the-corporate-landlord/.

Robertson, Adi. "OLPC's $100 Laptop Was Going to Change the World—Then It All Went Wrong." *Verge*, April 16, 2018. https://www.theverge.com/2018/4/16/17233946/olpcs-100-laptop-education-where-is-it-now.

Robinson, Arthur H. *The Look of Maps*. Madison: University of Wisconsin Press, 1952.

Rogers, Will. "Wall Street Journal." *Encyclopedia Britannica*. Accessed December 6, 2017. https://www.britannica.com/.

Rowe, Colin. *The Mathematics of the Ideal Villa and Other Essays*. Cambridge, MA: MIT Press, 2009.

Ruotolo, Kristin Jensen, Andrew Stauffer, Ivey Glendon, Jennifer Roper, and Kara McClurken. "Book Traces @ UVA." Report. University of Virginia, July 26, 2017. https://libraopen.lib.virginia.edu/public_view/cv43nw88q.

Safire, William. "Location, Location, Location." *New York Times*, June 26, 2009, Magazine sec. https://www.nytimes.com/2009/06/28/magazine/28FOB-onlanguage-t.html.

Schivelbusch, Wolfgang. *The Railway Journey: The Industrialization of Time and Space in the 19th Century*. Berkeley: University of California Press, 1986.

Shneiderman, Ben. "The Big Picture for Big Data: Visualization." *Science* 343, no. 6172 (February 14, 2014): 730. https://doi.org/10.1126/science.343.6172.730-a.

Schön, Donald A. *The Reflective Practitioner: How Professionals Think in Action*. New York: Basic Books, 1983.

Seaver, Nick. "Algorithms as Culture: Some Tactics for the Ethnography of Algorithmic Systems." *Big Data and Society* 4, no. 2 (December 1, 2017): 1–12.

Seaver, Nick. "Knowing Algorithms." In *Media in Transition* 8, 2013. http://nickseaver.net/s/seaverMiT8.pdf.

Seaver, Nick. "The Nice Thing about Context Is That Everyone Has It." *Media Culture & Society* 37, no. 7 (August 24, 2015): 1101–1109.

Sengers, Phoebe, Kirsten Boehner, Shay David, and Joseph "Jofish" Kaye. "Reflective Design." In *Proceedings of the 4th Decennial Conference on Critical Computing: Between Sense and Sensibility*, 49–58. New York: ACM, 2005.

Simmons, Joseph P., Leif D. Nelson, and Uri Simonsohn. "False-Positive Psychology: Undisclosed Flexibility in Data Collection and Analysis Allows Presenting Anything as Significant." *Psychological Science* 22, no. 11 (November 1, 2011): 1359–1366.

Simon, Herbert Alexander. *The Sciences of the Artificial*. Cambridge, MA: MIT Press, 1996.

Sismondo, Sergio. "Models, Simulations, and Their Objects." *Science in Context* 12, no. 2 (1999): 247–260.

Small, Bridget. *FTC Report Examines Data Brokers*. Federal Trade Commission, 2014. https://www.consumer.ftc.gov/blog/ftc-report-examines-data-brokers.

Smart Chicago Collaborative. "Crime and Punishment in Chicago." Accessed May 1, 2018. crime-punishment.smartchicagoapps.org/index.html.

Snopes. Accessed December 10, 2017. https://www.snopes.com/.

"Stanford CoreNLP—Natural Language Software | Stanford CoreNLP." Accessed April 23, 2018. https://stanfordnlp.github.io/CoreNLP/index.html.

Star, Susan Leigh, and James R. Griesemer. "Institutional Ecology, 'Translations,' and Boundary Objects: Amateurs and Professionals in Berkeley's Museum of Vertebrate Zoology, 1907–39." *Social Studies of Science* 19, no. 3 (1989): 387–420.

Star, Susan Leigh, and Karen Ruhleder. "Steps toward an Ecology of Infrastructure: Design and Access for Large Information Spaces." *Information Systems Research* 7, no. 1 (March 1, 1996): 111–134.

Steen, Francis, and Mark B. Turner. "Multimodal Construction Grammar." SSRN Scholarly Paper. Rochester, NY: Social Science Research Network, October 29, 2012.

Strauss, Valerie. “34 Problems with Standardized Tests.” *Washington Post*, April 19, 2017. https://www.washingtonpost.com/news/answer-sheet/wp/2017/04/19/34-problems-with-standardized-tests/.

Suchman, Lucille. *Human-Machine Reconfigurations: Plans and Situated Actions*. New York: Cambridge University Press, 2007.

Taub, Amanda, and Max Fisher. “Where Countries Are Tinderboxes and Facebook Is a Match.” *New York Times*, April 21, 2018, Asia Pacific sec. https://www.nytimes.com/2018/04/21/world/asia/facebook-sri-lanka-riots.html.

Taylor, Ann, Mitchell Marcus, and Beatrice Santorini. “The Penn Treebank: An Overview.” In *Treebanks: Building and Using Parsed Corpora*, edited by Anne Abeillé, 5–22. Dordrecht: Springer, 2003.

Thatcher, Jim, David O'Sullivan, and Dillon Mahmoudi. “Data Colonialism through Accumulation by Dispossession: New Metaphors for Daily Data.” *Environment and Planning D: Society and Space* 34, no. 6 (December 1, 2016): 990–1006.

Thorp, Jer. “A Sort of Joy.” July 28, 2015. https://medium.com/memo-random/a-sort-of-joy-1d9d5ff02ac9#.kcpo7fvlh.

Timberg, Craig. “Russian Propaganda May Have Been Shared Hundreds of Millions of Times, New Research Says.” *Washington Post*, October 5, 2017, Switch sec. https://www.washingtonpost.com/news/the-switch/wp/2017/10/05/russian-propaganda-may-have-been-shared-hundreds-of-millions-of-times-new-research-says/.

Tonnies, Ferdinand, and Charles Price Loomis. *Community and Society*. North Chelmsford, MA: Courier Corporation, 1957.

Totty, Michael. “The Rise of the Smart City.” *Wall Street Journal*, April 16, 2017. https://www.wsj.com/articles/the-rise-of-the-smart-city-1492395120.

Townsend, Anthony. *Smart Cities: Big Data, Civic Hackers, and the Quest for a New Utopia*. New York: W. W. Norton and Company, 2014.

“Treebank-3–Linguistic Data Consortium.” Accessed May 14, 2018. https://catalog.ldc.upenn.edu/ldc99t42.

Tronto, Joan C. *Moral Boundaries: A Political Argument for an Ethic of Care*. London: Psychology Press, 1993.

“Trump to CNN Reporter: You Are Fake News.” CNBC, January 11, 2017. https://www.cnbc.com/video/2017/01/11/trump-to-cnn-reporter-you-are-fake-news.html.

Tufte, Edward R. *The Visual Display of Quantitative Information*. 2nd ed. Cheshire, UK: Graphics Press, 1983.

Turkle, Sherry. *Life on the Screen: Identity in the Age of the Internet*. New York: Simon and Schuster, 1995.

Turkle, Sherry. *Simulation and Its Discontents*. Cambridge, MA: MIT Press, 2009.

Turkle, Sherry, and Seymour Papert. “Epistemological Pluralism: Styles and Voices within the Computer Culture.” *Signs* 16, no. 1 (1990): 128–157.

Turkle, Sherry, and Donald Schön. "The Athena Project." Massachusetts Institute of Technology, May 1988.

Turnbull, David. "Local Knowledge and Comparative Scientific Traditions." *Knowledge and Policy* 6, no. 3–4 (September 1, 1993): 29–54.

Turner, Fred. "Actor-Networking the News." *Social Epistemology* 19, no. 4 (2005): 321–324.

Uber. Accessed December 18, 2017. https://www.uber.com/.

"UCLA Library NewsScape." Accessed May 16, 2018. http://tvnews.library.ucla.edu/.

University of California at Berkeley, College of Environmental Design. "The Data Made Me Do It." March 13, 2015. https://ced.berkeley.edu/events-media/events/2015-studio-one-symposium.

University of Texas at Austin, Texas Advanced Computing Center. "The Future of Search Engines: Researchers Combine Artificial Intelligence, Crowdsourcing, and Supercomputers to Develop Better, and More Reasoned, Information Extraction and Classification Methods." *ScienceDaily*. Accessed December 7, 2017. https://www.sciencedaily.com/releases/2017/08/170803100019.htm.

Vertesi, Janet, and Paul Dourish. "The Value of Data: Considering the Context of Production in Data Economies." In *Proceedings of the ACM 2011 Conference on Computer Supported Cooperative Work*, 533–542. New York: ACM, 2011.

Vertesi, Janet, David Ribes, Laura Forlano, Yanni Alexander Loukissas, and Marisa Leavitt Cohn. "Engaging, Designing, and Making Digital Systems." In *The Handbook of Science and Technology Studies*, edited by Ulrike Felt, Rayvon Fouché, Clark A. Miller, and Laurel Smith-Doerr, 169–194. 4th ed. Cambridge, MA: MIT Press, 2016.

Victor, Daniel. "Trump's Victory, on Front Pages Worldwide." *New York Times*, November 9, 2016, Business Day sec. https://www.nytimes.com/2016/11/10/business/media/trumps-victory-on-front-pages-worldwide.html.

Wagner, Alex. "The Americans Our Government Won't Count." *New York Times*, April 2, 2018, Opinion sec. https://www.nytimes.com/2018/03/30/opinion/sunday/united-states-census.html.

Wall Street Journal. "2016 Presidential Election Calendar." September 10, 2015. http://graphics.wsj.com/elections/2016/calendar/.

Watson-Verran, Helen, and David Turnbull. "Knowledge Systems as Assemblages of Local Knowledge." In *Handbook of Science and Technology Studies*, edited by Sheila Jasanoff, Gerald E. Markle, James C. Peterson, and Trevor Pinch, 115–139. Thousand Oaks, CA: SAGE Publications, 1995.

Watzman, Nancy. "Internet Archive TV News Lab: Introducing Face-O-Matic, Experimental Slack Alert System Tracking Trump, and Congressional Leaders on TV News." *Internet Archive Blogs* (blog), July 19, 2017. https://blog.archive.org/2017/07/19/introducing-face-o-matic/.

"Wayback Machine." June 14, 1997. https://web.archive.org/web/19970614160127/http://www.cis.upenn.edu:80/~treebank.

Weizenbaum, Joseph. "Computer Power and Human Reason." In *The New Media Reader*, edited by Noah Wardrip-Fruin and Nick Montfort, 367–376. Cambridge, MA: MIT Press, 2003.

Wilson, Matthew W. *New Lines: Critical GIS and the Trouble of the Map*. Minneapolis: University of Minnesota Press, 2017.

Wood, Denis, and and John Fels. *The Natures of Maps: Cartographic Constructions of the Natural World*. Chicago: University of Chicago Press, 2008.

Wyer, Mary, Mary Barbercheck, Donna Cookmeyer, Hatice Ozturk, and Marta L. Wayne, eds. *Women, Science, and Technology: A Reader in Feminist Science Studies*. 2nd ed. New York: Routledge, 2008.

Yelp. Accessed December 18, 2017. https://www.yelp.com/.

"Zillow: Real Estate, Apartments, Mortgages, and Home Values." Accessed December 18, 2017. http://www.zillow.com.

Zimmerman, Ann S. "New Knowledge from Old Data: The Role of Standards in the Sharing and Reuse of Ecological Data." *Science, Technology, and Human Values* 33, no. 5 (September 1, 2008): 631–652.

Zuckerman, Ethan. *Digital Cosmopolitans: Why We Think the Internet Connects Us, Why It Doesn't, and How to Rewire It*. New York: W. W. Norton and Company, 2015.

Zukin, Sharon, Scarlett Lindeman, and Laurie Hurson. "The Omnivore's Neighborhood? Online Restaurant Reviews, Race, and Gentrification." *Journal of Consumer Culture* 17, no. 3 (October 14, 2015): 459–479.

INDEX

Page numbers with *f* refer to figures.

Advertising, local in, 22
Agnosticism, place, 9
Agre, Phil, 162
Ailanthus altissima (tree of heaven), 27, 34, 123
Aldrin, Buzz, 6
Algorithms. *See also* NewsSpeak algorithm
  challenging normative, 175, 176*f*, 177*f*, 178
  gaming, 191–192
  natural language processing, 104, 191
  of oppression, 191
Algorithms-data link
  in news analysis, 1, 3, 104–105, 118, 121, 191. *See also* NewsSpeak algorithm
  realities supported, 117–121
Ali, Christopher, 20
Allison, Leanne, 164
Ananny, Mike, 9
Antidata, 91
Anti-Eviction Mapping Project, 162, 183, 184–185*f*, 186
Anxiety, data inspiring, 16
Appfest, 55, 56*f*, 57, 59–60
Armstrong, Neil, 6
Arnold Arboretum, 29, 31
  Bussey Brook Meadow, 27, 28*f*, 29, 34
  community connections, 34
  Explorer's Garden, 33
  Hemlock Hill, 34
  knowing, ways of, 19
  map, 30*f*, 36*f*
Arnold Arboretum, accessions records
  all fields used, 37–38
  formats, 38–39, 38*f*, 39*f*
  users of, 22
  value, 39
Arnold Arboretum accessions records, place and
  collector addresses, 50, 51*f*, 52
  as history, 40–41, 42–45*f*, 47, 48–50
  Prunus sargentii, 1, 31–32
  Torreya grandis, 32–34
  Tsuga Canadensis, 34–35, 37
  Tsuga Caroliniana, 34–35, 37
  visualization (revealing alternative conceptions of), 39–41, 47, 50, 52
Arnold Arboretum data
  creation practices, 32
  encounters with, 32
  global-local coexistence, 22
  limitations, 34
  role of, 34–35
  tag diagram, 33–34, 33*f*
  taxonomy, 32
  understanding, 15
Arnold Arboretum data, place in
  attachments to, 3, 189
  experiencing, 189–190
  knowledge and, 35, 37
  provenance as, 15, 32
Artificial intelligence, 191
Associative trails, 90
"As We May Think" (Bush), 89
Atlanta
  Eastside, 141
  housing data, visual context, 142–150
  real estate, data in the culture of, 123–127
  segregation, 142
  wealth gap, 142
Atlanta, Old Fourth Ward changes
  gentrification, 152
  map, 143*f*, 151
  property value timelines, 144–150*f*
  visualizations, 142
Atlanta housing
  fair market value, establishing, 153
  gentrification, 152, 154, 156–157*f*
  grassroots organizations, 143
  renter's crisis, 154
  social history, 142
Autocomplete, 191

Bannon, Steve, 108
Battles, Matthew, 59
Baudrillard, Jean, 118

Bear 71 (Mendes & Allison), 161, 164–166, 183, 186
Being Digital (Negroponte), 9
Bellacasa, Maria Puig de la, 194
Benedikt, Michael, 20
Berners-Lee, Tim, 9–10
Big data, 15–16, 52–53, 69
Big Short, The (Lewis), 5
Bogost, Ian, 140
Book duplicates, relevance of, 60
"Book Traces" project, 60, 61*f*, 62
Borges, Jorge Luis, 118
Borgman, Christine, 17
Bowker, Geoffrey, 7, 21, 72, 194
Brand, Stewart, 131
Breitbart News Network (BNN), 100, 102–103, 108–109
Bricolage, 59
Bridge, data as, 190
Brown, Vincent, 179, 182
Buell, Lawrence, 52
Burry, Mike, 5–6
Bush, Vannevar, 89–90
Bussey Brook Meadow, Jamaica Plain, 27, 28*f*, 29, 34, 175

Cable News Network (CNN), 95, 100, 108–109
Cadwalladr, Carole, 191
Capta, data versus, 13
Care
  data work as, 23
  defined, 23
  ethics of, 9
Castells, Manuel, 20
Cataloging practices, 1, 68
Chan, Anita, 10
Chattanooga Public Library, 53
Clinton, Hillary, 95, 112, 190
Closed captioning, 106–107
Community, housing and, 141–142, 162, 183, 184–185*f*, 186
Computing, local in, 21
Constraints
  data, 14, 17–18, 67
  on form, Legos as, 164
Consumerism, 141
Context
  in data analysis, 139–140
  data and, 126–130
  housing market in, 126, 131, 142–150
Counterdata, 91
Crawford, Kate, 16
Cresswell, Tim, 19
Critical, term usage, 8–9
Critical data studies, 8
Critical design exhibitions, 162
Critical design interventions, 162
Critical reflection, defined, 162
Critical reflection models, purpose of, 163
Culler, Jonathan, 15
Cultural machines, 128
Cyberspace, 20

Dalton, Craig, 52
Darton, Robert, 91
Darwin, Charles, 41
Data. *See also* Arnold Arboretum data; Digital Public Library of America (DPLA) data; Housing data; Zillow data
  aggregate, limitations, 190
  bias and skewing effects, 2
  as bridge, 190
  capta versus, 13
  challenging the dominant uses of, 91
  constraints, 14, 17–18, 67
  creation of, 32, 164
  diversity, 14–15
  guiding future research, 192–196
  as indexes to local knowledge, 161–163
  interfaces recontextualize, 125
  large-scale, 21
  limitations, 190
  meaning of, 17
  news becoming, 103, 118–120
  publicly available, 3–4, 9–10, 130, 136–137, 155
  raw, 21
  realities supported by, 117–121
  recontextualization of, 126–127, 178–180, 180*f*, 181*f*, 182
  term usage, 13–14, 52
  understanding, 1, 189

universal, 69
unpacking, 189
value of, 90
Data, conceptions of
as autonomous, 15
as big, 15–16
embeddedness, 15
as more than rhetorical, 17
singularity, 13–14
universality, 14–15
Data, goals for
access, 190
analysis, 191–192
optimization, 192
orientation, 189–190
Data analysis, comparative approaches to, 169, 170–173*f*, 174–175
Data assemblage, 15
Databases, relational, 14
Data collections
concreteness, 4
historically, 5
online, 4–5
personal, 169, 170–173*f*, 174–175
Data dirt, 67
Data documentary, 164–166, 165*f*
Datafication, 103, 118–120
Data performance, 166–169
Data presentation, place in, 166–169
Data Revolution, The (Kitchin), 15
Data settings
conceptualizing, 166
from data sets to, 1
models for exploring, 164–166, 165*f*
Data Shapes, 82, 83–88*f*
Data studies, 7–8
Dates, conventions for writing, 1
Dear Data (Lupi & Posavec), 161, 169–175, 185
Del Tredici, Peter, 32–33, 35, 37, 39, 50
Design
critical reflection, enabling, 162
human-centered, 163–164
as rhetorical strategy, 162
Design of Everyday Things, The (Norman), 163
Digital media, place agnosticism, 9
Digital Public Library of America (DPLA), 56–57, 62
Appfest, 55, 56*f*, 57, 59–60
characterizations, 91
goal of, 57
invisible localities, 59–60
librarian–internet scholar conflicts, 23
local, visualizing the, 69, 72–73
search example, 62, 65*f*
temporalities, visualizing, 72–73, 74–81*f*
website main page, 62, 64*f*
white/black search, 68, 70–71*f*
Digital Public Library of America (DPLA) data
absences, revealing, 68, 70–71*f*
as big data, 16
classifications, 63
collection, local in, 3
constraints, 67
contributors, 1
duplicates/redundancies, 60
errors, decoding, 67
format diversity, 14, 16
limitations, 190
rituals shaping, 68–69
schemata, 63, 67
shape of, 82, 83–88*f*
understanding, 189
Digital Public Library of America (DPLA) Library Observatory, 58*f*, 59–60
Dosman, Michael, 41
Dourish, Paul, 14, 102, 128
Drucker, Johanna, 13
DuBois, W. E. B., 116
Dunne, Anthony, 162

Edwards, Paul, 16, 22, 35
Elites, digital, 10
ELIZA, 115
Ellis Act, 183
Emerson, Ralph Waldo, 116
Enrichment, 57
Entities becoming data, 17
Evidence, data as, 17
Explorer's Garden, 33

Fair market value, 152
Fake news, 20, 191
Fake News Challenge, 112
Fillmore, Charles, 114
Filter bubbles, 22–23
Finn, Ed, 125, 128
Fisher, Berenice, 23
Five-thirty-eight, 13
Flows, space of, 20
Frampton, Kenneth, 20
Friedman, William, 35

Gabrys, Jennifer, 166
Galloway, Alexander, 125
Geertz, Clifford, 8, 19, 21, 128–129
Gentrification, 152, 154, 156–157*f*
Geolocation, 19–21
Georgia, housing data, 153–154
Gibson, William, 20
Gillespie, Tarleton, 103
Gitelman, Lisa, 21, 69
Global, local coexisting with the, 22
Goldenberg, Sergio, 107
Google Earth, 175, 191
Grammar theory, 114
Grant, Cary, 116
Grassroots Mapping, 161, 175–178, 176–177*f*, 186

Hack/hackathon, 59
Halpern, Orit, 89–90
Hanan, Joshua, 123
Haraway, Donna, 40
Harding, Sandra, 19, 23
Hemlock Hill, 34
Hemlock woolly adelgid (*Adelges tsugae*), 34
Hepburn, Katherine, 116
Homestead exemption act, 153–154
Home values, context in determining, 126
Housing
  community and, 141–142, 162, 183, 184–185*f*, 186
  social history, 142
Housing crisis, 5–6, 126
Housing data
  Atlanta, GA, 142–150, 153–154
  civic approach to, 141, 153, 158
  consumerist approach to, 141, 158
  fair market value, establishing, 153
  optimization and contestation, 192
  property value timelines, Atlanta, 144–150*f*
  publicly available, 153
  recontextualizing, 126–127, 141–142
  scales of operation, 21
  visual context, 142–150
Housing Justice League, 143, 154
Housing market
  in context, 126, 131, 142–150
  Zillow and the, 1, 126–127, 141
Howell, David, 139
Human-machine communication, 6
Humphries, Stan, 130, 138, 139
Hyperlinks, 90

Identity, professional, 68
Indexes, 3, 161–163
Information, organizing, 191
Information extraction/retrieval, 114
Infrastructure inversion, 72
Infrastructures, data
  academic (*see* NewsScape)
  defined, 57
  industrial, for-profit (*see* Zillow)
  initiatives, 91
  motivations, 92
  place attachments in, 57
  placing, 89–91
  understanding, 189
  working ethically with, 92
Infrastructure studies, 22
Ingestion, 57
Inscriptions, 14–15
Interfaces
  causing friction, creating, 178–179, 180*f*, 181*f*, 182
  frictionless, 125, 130–132, 136–140
  potential, 126
  process elements, 125–126
  recontextualize data, 125
  term usage, 57
  universalizing, 125
Internet, 4–5, 9–10

Internet Archive, 112
Interoperability, 57
Interventions, 62

Jefferson, Thomas, 91
Johnson, May Helen, 154
Jokes, 116
Jones, Karen Späck, 113–114

King, Martin Luther, Jr., 142
Kitchin, Rob, 8, 15
Knowing, ways of, 19
Knowledge, data as indexes to local, 161–163
Knowledge discovery, 89
Knowledge ecology, 35, 37
Knowledge systems, 19, 35

Language of New Media (Manovich), 104
Latour, Bruno, 14, 15, 119
Lauriault, Tracy, 8, 15
Learning, situated, 178
Legos, 164
Lewis, Michael, 5
Librarians, 23
Licht, Jeffrey, 63
Local
  in advertising, 22
  in computing, 21
  in data, 67, 90
  dimensions of the, 19
  meaning of, 22
  news, 20–23, 104
  term usage, 4, 18–19
  visualizing the, 69, 72–73
Local, delimiting the
  geolocation in, 19–21
  global, coexisting with, 22
  as invisible, 23
  as lesser, 18–19
  scales of operation, 21
  an unquestionable good, 22–23
  as virtual, 20
Local data
  as community-based practice, 20
  redundancy of, 21
Localism, 20, 22
Local practice, models of
  data analysis, comparative approaches to, 169, 170–173*f*, 174–175
  data presentation, place in, 166–169
  data settings, looking at the, 164–166, 165*f*
  indexes to local knowledge, 161–163
  interfaces causing friction, creating, 178–179, 180*f*, 181*f*, 182
  normative algorithms, challenging by creating counterdata, 175, 176*f*, 177*f*, 178
  principles to practices, 163–164
  relationships, data in building, 183, 184–185*f*, 186
Local readings, 7–8
Longfellow, Henry Wadsworth, 61, 62
Looser, Devoney, 62
Lunar landing data, 6
Lupi, Giorgia, 169, 174

Ma, Thomas, 68
Manovich, Lev, 18, 104, 118–119
Maps, housing, 142–143
McArdle, Gavin, 15
McLuhan, Marshall, 18
McQuirter, Marya, 68
Media
  data as, 18
  data creating, 116
Media Cloud, 112
Melville, Herman, 116
Memex, the, 89–90
Mendes, Jeremy, 164
Mitchell, William, 20
Moore, Henry, 179
Mortgage bond market collapse/mortgage data, 5–6
Murray, Janet, 139–140

Narrative, 18
Natural language processing (NLP), 104–105, 107, 112–117
Natural Language Toolkit (NLTK), 105, 107–108, 111, 114–116
Negroponte, Nicholas, 9–10
News
  algorithmic turn in the analysis of, 121

- becoming data, 103–105, 118–120
- closed captioning, 106–107
- fake, 20, 23, 104, 112, 120–121
- local, 20–23, 104
- temporality of, 107

NewsScape, 100
- algorithms-data link, 1, 3, 118, 191
- changing temporality, 107

NewsSpeak algorithm
- creating the, 105–109
- data artifacts, 110–113
- election terms, 96–99*f*, 100, 103, 108
- observations, 109–110
- purpose of, 106

New York Public Library, 63
Nextdoor, 125
NLP (natural language processing) algorithms, 104, 191
Noble, Safiya, 191
Norman, Donald, 163
Norman door, 163–164
Nye, David, 92

Objects, contestational, 162
One Laptop per Child (OLPC) project, 9
Open data, local guides to, 194–195
Ortner, Sherry, 7

Page X, 112
Panorama/panoramic, 40
Pariser, Eli, 23
Penn Treebank, 116
P-hacking, 2
Photographs as data, 29
Place
- data and, 29, 31, 52–53
- in data presentation, 166–169, 189
- in data sets, 166–169
- interfaces causing friction, creating, 178–180, 180*f*, 181*f*, 182
- meaning of, 20
- term usage, 19

Polarization, 23
Pollock, Anne, 190
Posavec, Stefanie, 169, 174
Presidential election (2016), 20, 95, 96–101*f*, 100, 103, 190
Procedurality, 139–140
Prunus sargentii, 1, 31–32
Public Laboratory for Open Science and Technology (Public Lab), 27, 175

Raby, Fiona, 162
Racism, 22, 191
Rascoff, Spencer, 130, 138
Real estate, data in the culture of, 123–127
Real estate agents, 130–131
Reality, creating, 117–121
Reflection, critical, 9, 162
Relationships, data in building, 183, 184–185*f*, 186
Rhetoric, procedural, 139–140
Ruhleder, Karen, 22

Sargent, Charles Sprague, 1, 31–32, 41
Schemata, data, 63, 67
Schivelbusch, Wolfgang, 40
Schön, Donald, 162
Science, civic, 175
Science and technology studies, 19, 35, 119
Sensor data, 6
Settings, data. *See* Data settings
Sexism, 191
SHRDLU project, 113–114
Simon, Herbert, 162
Situated, 19, 29
"Slave Revolt in Jamaica, 1760–1761," 161, 178–182, 180–181*f*, 186
Small multiples, 151
Smart city, 190, 194
Social media, 4–5, 120, 125
Sorting Things Out (Bowker & Star), 7, 72
Sort of Joy, A, 161, 166–169, 167*f*, 186
Star, Susan Leigh, 7, 22, 72
Status, professional, 68
Stuff of Bits, The (Dourish), 14
Surgical data, 6

Technological sublime, 92
Television, local news on, 20
Temporalities, visualizing, 72–73, 74–81*f*

Thatcher, Jim, 52
Torreya grandis, 32–34
Transcription
closed captioning, 106–107
on-the-fly, 111
Tronto, Joan, 23
Trump, Donald, 95, 100–102, 104, 108, 111, 112, 190
Tsuga canadensis, 34–35, 37
Tsuga caroliniana, 34–35, 37
Tufte, Edward, 8
Turkle, Sherry, 59
Twain, Mark, 116

Uber, 125
Universalism, digital, 9–11, 125

Visual Display of Quantitave Information, The (Tufte), 8
Visualizations
Dear Data postcards, 169, 170–173*f*, 174–175
of invisible structures, 82
of the local, 69, 72–73
shape of data, 82, 83–88*f*
Sort of Joy, A, 166–169
temporalities, 72–73, 74–81*f*
term usage, 8

Wall Street Journal (*WSJ*), 100–101, 103, 108–109, 115
Weizenbaum, Joseph, 115
"What We Talk about When We Talk about Context" (Dourish), 128
White supremacy, 22, 68
Wink data, 129–130
Winograd, Terry, 114
Winters, Niall, 9
World Wide Web, beginnings, 9–10

Yelp, 125

Zestimates, 138–140, 153
Zillow, 130
algorithmic interface, 137–140
data, contextualizing, 131
data, recontextualizing, 126–127
discursive interface, 136–137
gentrification, emphasizing, 152
housing market impacts, effect on, 141
interfaces, localized models, 189
map, 142–150
motto, 131
public data, use of, 130, 136–137, 155
tag line, 125
transparency, facilitating, 1
visual interface, 131–132, 133–135*f*
Zestimates, 138–140, 153
Zillow data
culture of real estate and, 141–142
interfaces recontextualize, 3, 141–142
operational function of, 17
publicly available, 136–137
recontextualizing data, 126–127